MIMESIS
INTERNATIONAL

PHILOSOPHY
n. 72

Veronica Neri

ETHICS AND THE ARTIFICIAL IMAGE

Accountability and Reliability for a New Status of the Visual

MIMESIS
INTERNATIONAL

The publication of this volume is financed by a grant from the Department of Informatics of the University of Pisa.

www.mimesisinternational.com
e-mail: info@mimesisinternational.com

Isbn: 9788869775000
Book series: *Philosophy,* n. 72

P.I. C.F. 02419370305

CONTENTS

To my parents, Tiziana and Giuseppe,
with love and gratitude

PREMISE

Generative artificial intelligence (GenAI) makes it necessary to write yet another chapter in the field of visual culture studies and, in particular, its ethical implications. In addition to producing texts and voices that approximate human ones, images of all kinds are also generated with GenAI: journalistic images, advertising, audio-video excerpts, illustrations, infographics, artwork, architectural designs, historical reconstructions, biomedical imaging, etc.

We flip through a newspaper and no longer know whether the image that fascinates, disturbs, or reassures us was created by a photographer's shot or by the algorithmic processes of GenAI. We look at a billboard on the street and are no longer able to recognize whether the testimonial looking at us straight in the eyes is real or whether it is just the algorithmic synthesis of idealized features. We see videos with people we know, and we don't know whether it was really them who performed certain actions. Moreover, mass media act as seamless propagators of these new visual entities.

This scenario raises new and very significant ethical reflections. These are due not so much to a new visual style the GenAI produces – after all, every epoch and every techno-artistic tool have produced a specific style over time – but to the relationships of trust (and credibility) that are established between organic subjects, artificial images, and GenAI systems. In other terms, such ethical considerations are triggered by the necessary reshaping of the concepts of truth, veracity, and deception, the possible discrimination and stereotyping in terms of gender, ethnicity, ideology, religious beliefs, etc., the emerging distortions of reality, and a new concept of co-authorship.

If we can assert the hypothesis of a new form of "aura", linked to the algorithmic process and a new interaction between a subjects and a technological system, the ease of access to these tools, use and dissemination speed of its products risk polarizing thinking and perception of a technologically shaped reality. Lack of awareness about pretermitted ethical implications can turn such risks into dangers, erasing any margin of possible human control and engendering a greater state of uncertainty.

Moreover, we are not only dealing with AI-generated images, but AI meanders through the web via increasingly refined machine learning processes, putting images in front of the viewer's eyes, influencing our preferences and choices.

Principles such as autonomy, beneficence, maleficence and justice, and, lastly, explicability such as suggested by Floridi – understood as the responsibility of those who create systems and those who use them, and as the explicability and transparency of the systems themselves – become inescapable ethical principles. To these, in accordance with Fabris's perspective, we add the principle of limit, a meta-principle to be understood both as an awareness of human limitations and as a recognition of the limitations inherent in machines. In light of this, what most compellingly calls for ethical reflection on artificial images, in my opinion, is the new communicative paradigm (and its limits) that GenAI is opening. How is it possible to open a common space when only one interlocutor is aware of such openness? We are not talking about machines with consciousness –although such an assertion depends on the hermeneutics of the semantics of the concept of consciousness – but we can say that currently images are generated without the system's understanding what they have generated and why certain signs appear rather than others in the generated image. In view of the preceding considerations, in this book, I shall seek to address principal ethical issues arising from the proliferation of images generated by AI in and through media.

Therefore, the first chapter aims to investigate the new *status* of the artificial visual and the new role assumed by artificially generated images also in relation to an imagination that we could define as algorithmic. In the second part, I reflect on the concept of prediction and cognitive distortion with (and through) artificial images until we reach specific cases peculiar to the mass media world of advertising and information, areas in which GenAI is increasingly used. The behavioral advertising, the personalization of even visual information shape thinking and alter the human being's freedom of choice. The image-generative AI in the visual information world aims more at the 'truth effect' than a more authentic truth, asking each person to take responsibility for what is published in respect of the right to inform oneself and be informed.

Finally, in the last chapter, the ethical principles underlying image GenAI are addressed, intersecting with recent European regulation [in particular the AI Act (2024) and, before that, the European Ethics Guidelines (2019)]. Transparency, (co)authorship and the potential infinity of a transformative and dialogic process distinguish a new visual dimension at the basis of which lies the linguistic capacity of the subject to generate the most appropriate question that will then be translated into image. Ethics by design and the redress by design collide with the vulnerability of the system and human subjects within a society centered on the control of risk, uncertainty, and the heuristics of fear. These are ultimately even more ontologically arbitrary images that risk leading us to the most radical nihilism, which we could define as ethical, and to the "de-facticization" of an increasingly simulacral reality. Are we already exposing ourselves to the risk of being confronted with a world made up of artificial *simulacra* (ingenerating relations between them), *à la* Baudrillard, that simulate object reality by daring to lose it permanently?[*,**]

---

* Expenditure fully financed under the PNRR project – PE00000013 named FAIR – Future Artificial Intelligence Research – M.4-C.2- I 1.3 – Notice 341/2022 CUP: I53C22001380006, funded by the European Commission under the NextGeneration EU programme. The views and opinions expressed in this document are solely those of the authors and do not necessarily reflect the official position of the European Union or the European Commission. Neither the European Union nor the European Commission can be held responsible for them. The language revision was financed within the framework of the PNRR project- Decreto Direttoriale no. 1409 of 14 September 2022 - Mission 4 "Istruzione e Ricerca" and, in particular C2 –1.1, PRIN project - PRIN PNRR P2022MFWB9 I-Well - Integral WELLbeing for human flourishing: healthecare and food from an ethical and legal perspective – CUP 153D23007310001.

** I would like to thank the coordinators of Spoke 1 on Human-Centered AI PNRR FAIR - Future Artificial Intelligence Research (in the context of which this research was conducted), in particular Dino Pedreschi and Francesco Marcelloni, as well as the members of Research Group 1.6, *Co-design Methodologies for trustworthiness-by-design* with special reference to Adriano Fabris, Anna Monreale and, Silvia Dadà. I would like to thank also the members of the *Good-AI Lab*, especially Giovanna Marotta and Alessandro Lenci and all the scholars who generously contributed valuable insights, such as Massimo Bergamasco, Marco Menon, Francesco Borghesi, Edoardo Ghezzani, Elio Grande, and Emanuele Fulvio Perri. I wish to extend my gratitude to Antonio Da Re and Francesca Marin for the stimulating reflections that emerged during the Padua Summer School on AI and Humanism.

CHAPTER 1
THE TECHNO-ARTIFICIAL IMAGE

1.1. *The Visual Turn and New Ethical Scenarios*

Over the centuries and in various disciplines, the term 'image' has taken on a plurality of meanings and nuances, both semantically and aesthetically as well as ethically. In the contemporary age, the image has been the object (and at the same time the subject) of radical transformations, in parallel with the introduction and, above all, with the spread of new techniques and technologies of visual communication. This started at the end of the nineteenth century – with photography and then cinema – and continues today following the digital technology and the emergence of artificial intelligence (AI).[1]

The new centrality of image, which has led some scholars to theorise the concept of iconosphere,[2] goes beyond that acquired

1 The first attempts at photography date back to Nicéphore Niépce (1826), and were perfected by Louis-Jacques Daguerre (1837), becoming the prototype of the industrialisation of the seriality of the image. With the projection by the Lumière brothers in Paris, at the *Salon Indien du Grand Café* (1895), and with the patenting of the kinetoscope by Thomas Edison (1889) in the United States, the 'new spectacle' of the moving image, the cinema, was born. See F. Casetti, *Eye of the Century: Film, Experience, Modernity*, Columbia University Press, New York 2008.

2 Jean Wahl first introduced the concept of the "iconosphere" (with the term "phonosphere") in 1955. By this term, he means a dimension specific to the age of mass communication, in which human beings are perpetually immersed in images that not only surround them but also mediate their perception of reality. See M. Di Monte, "Iconosphere," in «International Lexicon of Aesthetics», 1, 2018; J. Wahl, *La culture est-elle en péril? Débat sur ses moyens de diffusion: presse, cinéma, radio, télévision, Rencontres Internationales de Genève*, Editions de la Baconnière,

in the age of the video-sphere, as canonised by Régis Debray, in which "our visual is in constant rotation, pure rhythm, obsessed with speed".[3]

The aspect of representation, which in the past was characteristic of the image, today gives way to the one of expression, which is detached from objective reality.

There is more emphasis on the subjectivity and arbitrariness of the visual, both on the part of those who create or co-create it (with the help of an algorithmic machine, for example), and on the part of those who view, observe and interpret it. The advent of the techno-sphere has afforded the image an unexpected and indeed revolutionary autonomy.

Once again in the history of media, the strength and power of the visual was confirmed in the 1990s by a new discipline, the so-called *visual culture*.[4] This revival of iconocracy has been supported above

---

Neuchâtel 1955, p. 335. On the contrary, the term 'infosphere', first theorised by Boulding on 1970 "consists of inputs and outputs of conversation, books, television, radio, speeches, church services, classes, and lectures as well as information received from the physical world by personal observation [...]. It can be argued that development of any kind [...] is primarily dependent on a network of information flows". Cf. K.E. Boulding, *Economics as a Science*, McGraw-Hill, New York 1970, p. 16; A. Toffler, *The Third Wave*, Bantam, New York 1991, pp. 15-6. On the infosphere see the more recent: L. Floridi, "Ethics in the Infosphere," in «The Philosophers' Magazine», 16, 2001, pp. 18-9; Id., *The Fourth Revolution: How the Infosphere is Reshaping Human Reality*, Oxford University Press, Oxford 2014.

3 The videosphere is the last of the three mediological caesuras theorised by Debray (in *Vie et mort de l'image*, Éditions Gallimard, Paris 1992, pp. 283 ff; Id., *Media Manifestos*, Engl. transl. by E. Rauth, Verso, London 1996). It is immediately preceded by the graphosphere (focused on aesthetics, where the image is a representation, but also a physical object with an author, the artist, and an audience) and, even earlier, by the logosphere (or era of idols, from the invention of writing to the printing press, where the image is considered magical and salvific).

4 Cf. N. Mirzoeff, *An Introduction to Visual Culture*, Routledge, London 1999; C. Jenks (eds.), *Visual Culture*, London, Routledge 1995; see also A. Pinotti and A. Somaini, *Cultura visuale. Immagini, sguardi, media e dispositivi*, Einaudi, Torino 2016; Ž. Paić and K. Purgar (eds.), *Theorizing Images*, Cambridge Scolars Publishing, Newcastle 2016; A. Fabris,

all by technological change, which has given rise to the seamless circulation of digital images.[5] Such a process of exaltation *tout court* of technologies has led some thinkers, from Gunther Anders to Hannah Arendt, to the lucid observations of Vilém Flusser, to doubt the increasingly less human character of our choices and decisions in favour of an increasing role of apparatuses of a "non-human character".[6]

It is in this context that the *pictorial turn* begins. This includes the pictorial turn theorised by William J.T. Mitchell in the Anglo-American sphere, following certain reflections on Erwin Panofsky and Jonathan Crary, and the *iconic turn* systematised by Gottfried Boehm [7] in the European sphere, reflecting on the role of the visual

---

Aesthetics and Ethics in Technological Environments, in Ž. Paić (ed.), *The Technosphere as a New Aestetic*, Cambridge Scholar Publishing, NewCastle 2022, pp. 74-88.

5 Some 'germs' relating to visual culture are already clearly present in Rudolph Arnheim in *Art and Visual Perception. A Psychology of the Creative Eye*, University of California Press, Berkeley and Los Angeles, 1954, and in *Visual Thinking*, University of California Press, Berkeley-Los Angeles 1969, but also in Aby Warbug's *Frammenti sull'espressione* (edited by S. Müller, It. transl. by M. Ghelardi and G. Targia, Edizioni della Normale, Pisa 2011) in which short writings dating from 1888 to 1905 are collected, in the cinematic visual culture studied by Béla Balàzs (*Der sichtbare Mensch oder die Kultur des Films*, Deutsch-Österreichischer Verlag, Wien 1924) and, later, in Wittgenstein's thought (or. *Philosophische Untersuchungen*, eds. G.E.M. Anscombe and R. Rhees, Blackwell, Oxford 1953) and in Michael Baxandall (*Painting and Experience in Fifteenth Century Italy*, Clarendon Press, Oxford 1972).

6 V. Flusser, *Medienkultur*, ed. by S. Bollmann, Fischer Taschenbuch, Frankfurt am Main 1997.

7 W.J.T. Mitchell, *Picture Theory: Essays on Verbal and Visual Representation*, University of Chicago Press, Chicago 1994; Id., *Pictorial Turn: eine Antwort*, in H. Belting (ed.), *Bilderfragen. Die Bildwissenschaften im Aufbruch*, Fink, München 2002, pp. 37-48; J. Crary, *Techniques of the Observer. On Vision and Modernity in the Nineteenth Century*, MIT Press, Cambridge Mass. 1990; E. Panofsky, *Perspective as Symbolic Form* (1927), Engl. transl. by C. Wood, Zone Books, New York 1991; G. Boehm, *Die Wiederkehr der Bilder*, in Id. (ed.), *Was ist ein Bild?*, Fink, München 1994, pp. 11-38; G. Boehm and W.J.T. Mitchell, "Pictorial Versus Iconic Turn: Two Letters," in «Culture,

in Gadamer.[8] Both scholars, also starting from common foundations such as the work of Nelson Goodman and Charles Sanders Peirce,[9] aim to overcome Rorty's *linguistic turn*, according to Vico's logic of historical recurrence that has characterised the history of the image over the centuries. Rather than a real 'turn', the interest in the visual is in fact a *topos* that periodically appears in philosophical thought: sometimes it is considered to represent the most refined achievements of a certain civilisation, while sometimes its most repulsive side, to be condemned. According to Rorty, words are the means of access to knowledge and reality.

> [A]ll linguistic philosophers talk about the world by means of talking about a suitable language. This is the linguistic turn, the fundamental gambit as to method, on which ordinary and ideal language philosophers (OLP, ILP) agree. Equally fundamentally, they disagree on what is in this sense a 'language' and what makes it 'suitable'.[10]

In order to identify a philosophical current that demonstrates the role of language as the foundation of any argument, Rorty introduces the concept of the linguistic turn. This term is used to describe the idea that language can be considered an autonomous form of life, with its own set of rules and conventions. In Rorty's view, thought is oriented towards the promotion of dialogue and listening. He asserts that the foundation of interactions between humans and reality lies not so much in images or representation in a general sense, but in the communicative (linguistic) process, which implies a continuous

---

Theory and Critique», 50, 2009, pp. 2-3; H. Belting, *Likeness and Presence. A History of Image before the Era of Art*, The Chicago University Press, Chicago 1997; G. Böhme, *Theorie des Bildes*, Fink, München 1999.

8 H.G. Gadamer, *Wort und Bild – 'so wahr, so seiend'* (1992), in Id., *Gesammelte Werke*, 10 vols., Tübingen, Mohr, 1985-1995, vol. 8 (1993), pp. 373-99.

9 In this regard, see W.J.T. Mitchell, "Pictorial turn. Una risposta," in «Lebenswelt», 2, 2012, pp. 130-42.

10 R. Rorty, *Introduction. Metaphilosophical Difficulties of Linguistic Philosophy*, in Id. (ed.), *The Linguistic Turn: Recent Essays in Philosophical Method*, University of Chicago Press, Chicago 1967, p. 8.

exchange within the framework of an ethically and socially oriented action.[11]

Mitchell, for his part, believes that a paradigm shift is occurring in which vision is as significant as language in mediation of social relations.

> [A] postlinguistic, postsemiotic rediscovery of the picture as a complex interplay between visuality, apparatus, institutions, discourse, bodies, and figurality. It is the realization that spectatorship (the look, the gaze, the glance, the practices of observation, surveillance, and visual pleasure) may be as deep a problem as various forms of reading (decipherment, decoding, interpretation, etc.) and that visual experience or "visual literacy" might not be fully explicable on the model of textuality. Most important, it is the realization that while the problem of pictorial representation has always been with us, it presses inescapably now, and with unprecedented force, on every level of culture, from the most refined philosophical speculations to the most vulgar productions of the mass media. Traditional strategies of containment no longer seem adequate, and the need for a global critique of visual culture seems inescapable.[12]

The iconic turn, then, is an expression by which one recognises in the image both an object made, consisting of concrete and tangible matter, bits and/or algorithmic calculations, and an entity to be interpreted that transcends its 'materiality'. Belting's research seems to go in the same direction. He emphasises the polyvocal dimension of the image and the medium – even more so with digital, network and AI – that makes it visible and characterises it.[13]

In light of the aforementioned considerations, it seems pertinent to examine the ethical implications of the contemporary image in a technological context that is characterised by rapid evolution and significant differences from that of a few years ago. The centrality of the visual paradigm and the advent of the *algorithmic turn* demand

11 Id., *Contingency, Irony, and Solidarity*, Cambridge University Press, Cambridge 1989, pp. 153-4 ff.

12 W.J.T. Mitchell, *Picture Theory*, cit., p. 16.

13 H. Belting, "Image, Medium, Body: A New Approach to Iconology," in «Critical Inquiry», 31, 2, 2005, pp. 73 ff.

further reflection on the ontological and ethical condition of the image.[14] Unlike more traditional computer graphics programs, the image can be reproduced and modified without interruption, yet it is generated by algorithms that act autonomously with respect to the will and desires of the subject. As a result, the individual cannot predict or control the outcome of the image-generating system in detail.

In the context of the technical reproducibility of analogue and, even more so, digital photography, the image is effectively deprived of its "aura". This phenomenon can be elucidated by referring to Walter Benjamin's paradigmatic and revolutionary essay, *The Work of Art in the Age of Its Technological Reproducibility*.[15] Conversely, with the advent of AI, a unique entity is created with each instance of image generation. Subsequently, the image becomes reproducible in the same manner as any other digital image or photograph. Is this perhaps a novel form of "aura"?

Benjamin read the technical reproducibility of the image (alluding in particular to the photographic image) as a move towards a clear break with the aristocratic concept of the image, the desacralisation of the elite image and the destruction of the "aura", the so-called *hic et nunc* of the work of art. He observes new modes of reception of the work of art and a new perception of the world culturally and historically situated. With image-generating artificial intelligence, as we shall see, we can perhaps still speak of artistic images,

---

14 W. Uricchio, "The Algorithmic Turn: Photosynth, Augmented Reality and the State of the Image," in «Visual Studies», 1, 2011, pp. 25-35; G. Bolin (ed.), *Cultural Technologies: The Shaping of Culture in Media and Society*, Routledge, New York 2012.

15 The work had five different versions, four in German and one in French, written between 1935 and 1939, all unfinished. The French version, translated by Pierre Klossowski, is the only one published by the author during his lifetime and was subsequently rejected because it had undergone some cuts without his prior consent, giving an altered view of his thought. See W. Benjamin, *The Work of Art in the Age of Its Technological Reproducibillity* (1936, second version), in Id.,*The Work of Art in the Age of Its Technological Reproducibility, and Other Writings on Media*, Engl. transl. by M. W. Jennings *et al.* and eds. by E. Jephcott *et al.*, The Belknap Press of Harvard University Press, Cambridge Mass. London 2008.

individualistic-aristocratic *à la* Benjamin, but also of a mass image, more popular, within reach even of those without specific technical skills. In this way, a new perception of the world and of 'another world' certainly emerges, but also a new concept of *hic et nunc*, this time not entirely of human origin. This binary tendency of the image thus seems to be present in generative AI (so-called GenAI), where the growth of technological and mass-media potential has led to a multiplication of the ways in which images can be created and perceived.

Benjamin argues that the medium is a way in which human sensory perception is systematized, while *Apparat* refers to the technical media of modernity through which 'things' manifest themselves in an environment that directs their meanings and enables their interpretation. Indeed, as Somaini writes:

> If, on the one hand, the term *Apparat* (which becomes *Apparate* in the plural, and *Apparatur* if taken as a collective singular) indicates in Benjamin the various material and technical artifacts – architectural constructions, technical instruments, sound and optical devices, means of (mass) communication – that contribute, through their various operations, to the organisation of the field in which sensory experience takes place, the term *Medium* (a term that I will always write with the German spelling, with the capital "M" and in italics, whenever I refer specifically to its meanings in Benjamin's writings and in German literature) precisely indicates such a field: the spatially extended environment, the *milieu*, the *Umwelt* in which perception occurs. What we may consider as Benjamin's "media theory" is therefore the study of the interplay between the historically-evolving domain of the technical *Apparate* and the "Medium of perception" in which they operate, determining different "articulations of the real".[16]

It can thus be argued that image-generating artificial intelligence models represent both the *Apparat* and the *Medium* in Benjamin's terms. Algorithmic systems can be seen to function as the means through which images are processed. While they do not afford us

16 A. Somaini, *The "Medium of Perception: Walter Benjamin's Media Theory and the Tradition of The Media Diaphana"*, in Ž. Paić and K. Purgar (eds.), *Theorizing Images*, cit., pp. 86-7.

complete freedom of choice (except within the framework of artificial intelligence proposals), they also constitute the environment in which their perception takes place,opening up new ethical scenarios for action.

1.2. *The Image Between Natural and Artificial*

In the techno-sphere populated by new media and new *Apparatuses*, images, as Flusser argued as early as 1978, no longer have the role of merely representing the world. They make it possible to give meaning to theories about the world. Alphabetic codes make it possible to describe concepts conveyed by images; techno-codes, on the other hand, transform concepts into images. And, in visual form, these concepts (un)consciously influence the actions of individuals.[17] After all, "[t]he image does not reproduce, but manifests, shows".[18]

From an ethical point of view, therefore, the question is not so much what images show and how, but the values behind their creation and what effect images (and the media through which they are manifested) have on people's choices and decisions as part of a society and/or as individuals. That is, from an ethical point of view, to what extent do images 'move' our minds and souls by suggesting that we consciously or unconsciously follow certain directions?

To address this question, it is first essential to examine the term image.[19]

---

17 V. Flusser, "*Die kodifizierte Welt,*" in «Merkur: Deutsche Zeitschrift für europäisches Denken», 359, 32, 4, 1978.

18 A. Ferrarin, *Un mondo non di questo mondo*, Edizioni ETS, Pisa 2023, p. 170.

19 I will limit myself here to a brief etymological reference. For a more comprehensive and articulated theory of the image, see: J.-J. Wunenburger, *Philosophies des images*, PUF, Paris 1997; Id., *La vie des images* (1995), Les Éditions Spéciales, Paris 2022; V. Flusser, *Ins Universum der technischen Bilder*, European Photography, Göttingen 1985, Engl. transl. by N.A. Roth, *Into the Universe of Technical Images*, University of Minnesota Press, Minneapolis 2011; J. Aumont, *L'image*, Armand Colin, Paris 2005; G. Didi-Huberman, *Images malgré tout*, Minuit, Paris 2003;

The image is traditionally defined as such insofar as it is visible, in object reality or in our minds. It is always an 'image of', implying a duality between the object as it is perceived by the eye and the object's signified meaning. Two distinct semantic nuances emerge, namely an imitation – sometimes distorted, sometimes very similar – of the original, but also another identity to be shown. This duality also seems to be evoked by the doubling of the letter *b* in the German noun *Ab-bild* and the letter *m* in the Italian word *im-magine*.[20] At the moment of its technological reproducibility, this duality seems to be transformed into a kind of multiplicity of senses that follow one another, by virtue of the speed with which the image can be modified and can itself change once inside an apparatus intended *à la* Flusser.[21]

From an etymological perspective, the term image, "expresses a profound semantic ambiguity," as observed by Wunenburger. It is necessary to investigate the etymology of the term in the context of Neo-Latin languages. The word 'image', which encompasses the meanings of natural, artificial and mental, has its roots in the Greek lemma *èikon* (icon), derived from the Indo-European root **weik-*, which refers to that which is offered to the eye, and thus to the concept of resemblance.

The term *èikon* bears close relation to the terms *èidos* and *èidolon*, and also to the concept of unreality. The Greek root * *id-* ultimately evokes the verb *horào*, meaning 'to see', in and through the multiple dimensions of reality.[22]

The term image is derived from the Latin *imago-inis*, meaning a figurative representation, an external form of an object, perceived

---

A. Fabris, *Philosophy, Image and the Mirror of Machines*, in Ž. Paić and K. Purgar (eds.), *Theorizing images*, cit., pp. 111-20.

20 Cf. A. Ferrarin, *Un mondo non di questo mondo*, cit., pp. 163 ff; U. Perone, "Il nome e l'immagine: la differenza dell'identità," in «Teoria», 1, 2010, pp. 72-3.

21 On the difference between technique and technology, see A. Fabris, *Ethics of Information and Communication Technologies*, Springer, Cham 2018, pp. 2-6.

22 D.J. Chalmers, *Reality+: Virtual Worlds and the Problems of Philosophy*, Allen Lane, London 2022.

by sight, referring to the concept of sensible form. It is related to the adjective *aemulus*, 'emulating', and the verb *imitor*, 'to imitate'. That is, it indicates the imitation of something from concrete reality. But it also takes on the meaning of illusory apparition and vision, as can be seen in a passage from Virgil's *Aeneid*.[23]

To sum up, *imago* and *èikon* refer to two semantic registers: the first with the meaning of material representation, ascribable to the idea of visible form, which we could call external image; the second, on the other hand, as a mental representation, which is a kind of internal image,[24] ascribable to the idea of unreal content, associated with imagination.[25]

---

23 J.-J. Wunenburger, *Philosophie des images*, PUF, Paris 2001, p. 30. See also the headword *Imago*, in *il Latino*, dictionary edited by G.B. Conte, E. Pianezzola, G. Ranucci, Le Monnier, Milano 2010, p. 692; Virgil, *The Aeneid*, Engl. transl. by D. West, Penguin, London 2003, book 1, v. 353: "Ipsa sed in somnis inhumati venit imago/coniugis"; Headword *Imago-inis*, in F. Calonghi, *Dizionario latino-italiano*, Rosenberg & Sellier, Torino 1989, p. 1306.

24 This is also clear in the English language where 'image' means a generic image, especially a mental one, while 'picture' means a material image.

25 For an analysis of internal, mental and mnestic images, see A. Ferrarin, *Un mondo non di questo mondo*, cit., pp. 180-218. According to Flusser, however, this distinction is arguably superfluous, given that mental images, which are internal, and material images, which are external, are in fact complementary: you need to create (or imagine) one in order to create (or imagine) the other. Cf. F. Restuccia, *Il contrattacco delle immagini, tecnica, media e idolatria da Vilém Flusser*, Meltemi, Milano 2021, pp. 247-8; M. Ferraris, *L'immaginazione*, Il Mulino, Bologna 1996. Furthermore, both registers are partially identified in the classification of images proposed by Wunenburger, which is based on three typologies. The first level encompasses sensible representations – immediate perceptual impressions and products of imagination – aligned with the Latin notion of *imago*. At the second level are images, according to the empiricist tradition, as abstractions grounded in sensory experience, extending to conceptual representations. Finally, there is the third, intermediate level, where image means an idea of representation of an absent reality, the perception of which persists in the absence of the representation itself, giving rise to mnestic or imaginative processes. These "ways of the imaginary," as previously outlined, are systematized by Wunenburger into three categories: *imagerie*, imaginal, and imaginary.

In light of this dualism, one cannot claim that the human gaze is inherently neutral or that there exists a "degree zero" of the gaze with respect to natural or artificial images. At most, an ethics of the gaze, far from AI, can be implied in the iconoclastic diatribe that the Council of Nicaea (787) tried to put an end to.[26] A gaze that is also twofold, that seeks to return the image to the known, and that at the same time allows itself to be surprised by the unknown, the transcendent or the magical, with new meanings.

In addition to this dualism, contemporary 'external' images can be classified into two principal categories: the natural and the so-called 'technical' ones.[27]

It can be assumed that natural images are those that are perceived by the retina of the eye when a particular object is observed.

Even natural images have an observer, and they do not exist without one; in fact, they cease to exist when the observer ceases to look at them. In this context, there is no issue with copying, reproduction, simulacra, or *mìmesis*, which are characteristics of other types of images.

Its reflection is inherent in the very concept of a natural image. When a natural element is reflected in water, for example, one has the reflected image and at the same time the thing itself. The idea and interpretation of this type of image depends on the awareness of the beholder. With experience, one learns to distinguish reality from its most common representation proposed by nature itself.

As far as technical images are concerned, it can be stated that their prevalence has increased disproportionately over time, to the extent that they now permeate our lives. It can be assumed that

---

These categories encompass objects from a variety of disciplines, including semiology, symbolic iconology, and a discipline of dream and fiction, which are seen to merge into one another. Cf. J.-J. Wunenburger, *La vie des images*, cit., p. 30 ff.

26 On the subject, cf. M. Bettetini, "The Status of Image: A Challenge for Medieval Aesthetics," in «Acta Philosophica», 2, 30, 2015, pp. 265-84.

27 On technical images, cf. V. Flusser, *Into the Universe*, cit., 144 ff.; on the difference between technical and natural images: A. Fabris, *Five Philosophy, Image and the Mirror of Machines*, in Paić, Ž. and Purgar, K. (eds.), *Theorizing Images*, Cambridge Scolars Publishing, Newcastle 2016, pp. 111-120.

there are two levels of technical images. The first level includes images produced using traditional and handicraft techniques, such as painting, watercolour, engraving, etching drawing, sculpture, and so forth. The second level comprises digital images, including photographs, digitised analogue images, and born-digital images. This category can now be expanded to include third-level technical images, which could be designated as 'artificial images'.[28]

By artificial images I do not mean all those that mankind has produced through the indiscriminate use of various techniques and technologies, that is produced with an *artificium*.[29] I am referring to the images produced by GenAI, which we will discuss in the next section. In the following section, I will focus on second-level technical images (which we will simply call 'technical images' from now on) and then, above all, on artificial images.

With regard to technical images, Flusser had visionary ideas for his time, long before the widespread use of the Web and in particular when virtual reality and the artificial image were still far from manifesting themselves in pictorial form. He did not distinguish technical images from natural ones, but rather from traditional ones: that is, he distinguished images on the basis of the technique or technology with which they were made: "The first arise through depiction, the second through a peculiar hallucinatory power that has lost its faith in rules".[30]

Images made with traditional techniques are thus images that result from our creative whim, conceived from emotions, experiences and dreams. Images made with technology, on the other hand, are considered to be more dependent on the apparatus that conveys them and the inherent limitations of the technology itself. An intuition that Flusser reinforces by describing a possible (and not too distant) future 'fable-like' scenario:

---

28 *Ibid.*; V. Flusser, *Into the Universe*, cit.; B.-C. Han, *Non-things. Upheaval in the Lifeworld*, Engl. transl. by D. Staeler, Polity Press, London 2022.

29 From the Latin *artificium*, from *artifex* 'maker, creator', meaning craft, but also skill, ability and talent. Cf. Headword *Artificium*, in G.B. Conte *et al.*, *il Latino*, cit., p. 128.

30 V. Flusser, *Into the Universe*, cit., p. 10.

> The scenario, the fable, I propose here is this: people will sit in separate cells, playing with their fingertips on keyboards, staring at tiny screens, receiving, changing and sending images. Behind their backs, robots will bring them things to maintain and reproduce their derelict bodies. People will be in contact with one another through their fingertips and so form a dialogical net, a global superbrain, whose function will be to calculate and compute improbable [and "probable", I would add] situations into pictures, to bring information, catastrophes about. Artificial intelligences will also be in dialogue with human beings, connected through cables and similar nerve strands. In terms of function, then, it will be meaningless to try to distinguish between natural and artificial intelligences (between primate brains and secondary brains).[31]

Flusser therefore upholds an assumption that is more valid today than ever, namely that the actuality of the image consists of hybrid images, in which- the traditional and the technical – and today we would add the technological – interact to the point where they are no longer distinguishable.

In light of the above, a technical image can be described as an image generated by an artificial apparatus, gradually asserting itself through algorithmic, pixel-based models that reconstruct a unified whole by means of point-like signs. These technical images represent not just efforts to visually describe the world but rather "consequences of scientific progress – a quest to depict the concepts the photographer holds of an image."[32] They unify and integrate scattered pieces, transitioning from "observations of objects" to "computations of concepts," that are multisensory and complex.[33]

The apparatus is what delimits the scope of action of the author or co-author of the image and the agency of the image. Unsurprisingly, Flusser has a very precise understanding of the apparatus. "The apparatus is not an extended thing. It is not something we can see,

31 *Ivi*, p. 161.

32 V. Flusser, *Kommunikologie*, hrsg. Von S. Bollmann und E. Flusser, Frankfurt a.M. 1998, p. 102. Cf. studies on Flusser by M. Menon, *Vilém Flusser e la «rivoluzione dell'informazione». Comunicazione, etica, politica*, Edizioni ETS, Pisa 2022, pp. 42 ff.

33 V. Flusser, *Into the Universe*, cit., p. 10.

touch, smell, or hear the hum of. What we see, touch, smell or hear humming is not the apparatus, but its instruments. The apparatus is what structures the instruments. We know that we are in front of an apparatus, or inside an apparatus, when all the things in front of us, or around us, are transformed into instruments. When, as if by magic, things cease to be in front of us, or in our hands, or to exist, and start functioning automatically".[34]

The subject (the "functionary" for Flusser) uses the *apparatus*. It represents "a new form of life, which differs from human life in that it does not 'exist', but functions, that is, behaves according to a programme," in a sense alienating the subject from the decision-making process. With the apparatus, therefore, an image is realised that unites and uniforms, with "uni-formation" meaning "the ability to recreate unity from the dispersion of the quantum universe".[35] Uni-formed surfaces: that is, surfaces in which minimal units are juxtaposed, aggregated according to a sense given to them. "The first are observations of objects [traditional images], the second computations of objects [technical images]".[36]

Flusser defines this ability to uni-form as the "capacity to step from the particle universe back into the concrete".[37] This unification involves several aspects and can take place between the representation of the image and its meaning, and between the image of an object and the object itself.

The apparatus and/or medium that enables the creation of a technical image, on the other hand, is nothing more than a kind of black box[38] that stands between the images and the viewer (or the user of the technical image). We do not know how the apparatus

34 Id., *O Último Juízo*, vol. II, È Realizações Editora, São Paulo 2017, pp. 326-27, cit. in M. Menon, *Vilém Flusser e la rivoluzione dell'informazione*, cit., p. 125.

35 *Ibid.*, pp. 132-9: 139.

36 V. Flusser, *Into the Universe*, cit., p. 10.

37 *Ibid.*, p. 34.

38 F. Di Pasquale, *The Black Box Society: The Secret Algorithms That Control Money and Information*, Harvard University Press, Cambridge Mass. 2016; D. Irrgang, *Vilém Flussers Black Box*, in E. Geitz, C. Vater, S. Zimmer-Merkle (eds.), *Interdisziplinare Perspektiven*, De Gruyter, Berlin-Boston 2020, pp. 53-69.

works, a structure within which every object or person becomes an instrument that functions automatically, according to a programme. "As if by magic, things cease to be in front of us, or in our hands, or to exist, and begin to function automatically".[39]

These Flusserian reflections on 'technical' images imply some ethical considerations, which we will focus on in this discussion. The first concern the second-level technical image, analogue and digitised or born-digital, disseminated on the network and 'managed' by algorithms through systems of machine and deep learning, chatbots, profiling (not only of one's own data, but also of one's own face and other visual information), nudges, filtering (collaborative and consumption-based filtering),[40] etc. We could also call this type of image a top-level artificial image. It is still an artificial image as opposed to a natural image.

These images, which can be manipulated and disseminated in real time, can be used to guide the individual's tastes (in terms of values, opinions, consumption choices, purchases, etc.) and decisions, but also to create *ad hoc* images. This is not a true process of persuasion, which is now almost obsolete, but a process of 'incorporating' the individual into the image (so-called 'customer engaging'), a performative practice in which the individual participates in the choice process through visual input, feeling embodied in the process itself. These are images that, for the first time, are produced in a virtual place and are transportable from one medium to another. They also appear more real than the object reality itself and aiming to homogenise the viewer's taste and reduce the sense of awe.[41]

In order to avert the de-responsibilisation of the subject, Flusser proposes to create images based on the dialogue with the machine, images produced without any apparatus, or 'silent' images.

'Silent' images ("*stille bilder*"), in particular, are able to outwit the device. They "force apparatuses to function against their peculiar progress," like, for example, those taken with a camera (analogue or

39 V. Flusser, *O Último Juízo*, cit., vol. II, pp. 326-7.

40 Cfr. I. Goodfellow *et al.*, *Deep Learning*, MIT Press, Cambridge Mass. 2016; R.H. Thaler and C.R. Sunstein, *Nudge. Improving Decisions about Health, Wealth, and Happiness*, Penguin Books, London 2011.

41 V. Flusser, *Bilderstatus*, in Id., *Medienkultur*, cit., pp. 69-82.

digital).[42] By playing against the apparatus, therefore, it is possible to lead it to produce something other than what it was programmed for. These images are meant to cover up "the flood of images", as a "defence and salvation from motion and sound images".[43] 'Silent' images are produced by transcending the apparatuses themselves, devices with which images are constructed from mathematical equations.

The images of the second digitisation introduce a deeper agency, they distance themselves more and more from our mind and body. And the act of perception becomes expanded, virtual, like gaze itself.[44]

Furthermore, more specific reflections, on the other hand, concern the image generated 'autonomously' by AI through more or less refined text strings (prompts); we are dealing with an image that could be described as third-level technical, or, more precisely, as 'artificial proper' or 'second-level artificial'.

As we will see in this discussion, these are images that force a rethinking of concepts of imagination (and, in some respects, of the imaginary) and risk generating new stereotypes and *biases* (or reinforcing those already consolidated over time, also on the basis of how the images in the dataset are labelled). Both are aspects that call for ethical inquiry, as we shall see later on.

1.3. *Generative Artificial Intelligence and Artificial Images*

No technology is born *ex abrupto*. And artificial images, created by GenAI, do not betray this assertion. Without going into detail about the origins of AI, which some scholars date back to the eighteenth century with Charles Babbage and Ada Lovelace, we can at least mention a few key moments in its history. In 1936, the mathematician Alan Turing came up with the idea of a machine,

---

42 *Ibid.*, p. 75.

43 *Ibid.*, p. 71.

44 I. Hoelzl and M. Marie, *Postimage*, in R. Braidotti and M. Hlavajova (eds.), *Posthuman Glossary*, Bloomsbury Academic, London-New York 2018, pp. 360-62.

known as the 'Turing machine', capable of reading data, performing calculations and writing down the results on an input/output tape.[45]

However, we had to wait until 1955, when four computer scientists (John McCarthy, Marvin L. Minsky, Nethaniel Rochester and Claude E. Shannon), first used the term 'artificial intelligence' and set up the Dartmouth Summer Research Project on Artificial Intelligence in 1956[46], an event that would mark the birth of the study of AI. In 1957, Rosenblatt theorised the perceptron[47], based on Hebb's studies, and two years later Samuel coined the expression 'machine learning',[48] to describe a program capable of developing algorithms to perform tasks

---

45 Thanks to Babbage (1791-1871) we have the first prototype of a mechanical computer, designed to perform generic operations. Lovelace studied these devices to perform complex algorithms for calculating specific numerical sequences. Cf. S. Henin, *Il computer dimenticato. Charles Babbage, Ada Lovelace e la ricerca della macchina perfetta*, edited by M. Temporelli, Hoepli Milano 2015; A.M. Turing, "Computing Machinery and Intelligence," in «Mind-New Series», 59, 236, 1950, pp. 433-60; M.A. Boden, *Artificial Intelligence: A Very Short Introduction*, Oxford University Press, Oxford 2018. With the so-called 'Turing test', an attempt was made to find out whether a machine 'could think'. If a human being is unable to distinguish whether the respondent is an individual or a machine, it can be concluded that the computer is exhibiting 'intelligent' behaviour. Cf. C. McKinstry, "Minimum Intelligent Signal Test: An Alternative Turing Test," in «Canadian Artificial Intelligence», 41, 1997, 35-47.

46 Cf. J. McCarthy *et al.*, "A Proposal for the Dartmouth Summer Research Project on Artificial Intelligence, August 31, 1955, in «AI Magazine», 27, 4, 2006, pp. 12-14.

47 Warren McCulloch and Walter Pitts propose that neurons represent the fundamental logical unit of brain activity. Cf. W.S. McCulloch and W. Pitts, "A logical calculus of the ideas immanent in nervous activity," in «Bulletin of Mathematical Biophysics», 5, 1943, pp. 115-33; F. Rosenblatt, "The Perceptron. A Perceiving and Recognizing Automaton", report no. 85-460-61, Cornell Aeronautical Laboratory Inc., January 1957.

48 A. Samuel, "Some Studies in Machine Learning Using the Game of Checkers," in «IBM Journal of Research and Development», 3, 3, 1959, pp. 210-29. In fact, Turing already addressed the concept of 'learning machines' in Id., "Computing Machinery and Intelligence", cit.; however, Samuel was the first to utilise the term in reference to a new field of computer science dedicated to machine learning.

without explicit instructions, relying instead on data-driven patterns and inference. Concurrently, an imagery of the artificial world emerges through literature and cinema. Consider, for instance, Isaac Asimov's literary works[49] or films such as *2001: A Space Odyssey* (1968) or the more recent *Ex Machina* (2015) and *Her* (2013). However, it is only in recent years that we can speak of the springtime of AI, thanks to the large amount of data now available that is sufficient to make algorithmic models work more reliably.

If anything, let us try to clarify the concept of intelligence, to see if it really applies to algorithmic systems. There is no single, unambiguous definition of intelligence. We can define it as the ability to accomplish complex tasks, adapting to the context; or, more precisely, "the ability to learn, understand, and make judgments or have opinions that are based on reason".[50] Just as there is no single, unambiguous definition of artificial intelligence. We can describe it as a procedure and, as such, it cannot be defined as natural intelligence, but rather as a somewhat approximate simulation of it. As Floridi states, it is rather "a new form of agency, not intelligence", a sort of "increasing reservoir of smart agency".[51] In 1955, McCarthy, Minsky, Rochester, and Shannon stated that "[...] the artificial intelligence problem is taken to be that of making a machine behave in ways that would be called intelligent if a human were so behaving".[52]

We can identify two different categories of AI, both are based on neural networks, the so-called *narrow AI* and *general AI.* The first describes all those systems that are designed and used to perform specific, albeit complex, tasks, such as, for example, playing chess,

49 I. Asimov, *I, Robot*, Gnome Press, New York 1950.

50 Headword *Intelligence* in Cambridge Dictionary online: https://dictionary.cambridge.org/dictionary/english/intelligence.

51 L. Floridi, *Ethics of Artificial Intelligence. Principles, Challenges, and Opportunities*, Oxford University Press, Oxford 2023, pp. 14, 6.

52 Cited by *Ibid.*, p. 35. Cf. https://www-formal.stanford.edu/jmc/history/dartmouth/dartmouth.html. See also the definition of AI in A. M. Kaplan and M. Haenlein, "Siri, Siri, in My Hand: Who's the Fairest in the Land? On the Interpretations, Illustrations, and Implications of Artificial Intelligence," in «Business Horizons», 62, 1, 2019, pp. 15-25 and in A. Fabris, *Filosofia nell'epoca dell'intelligenza artificiale*, Carocci, Roma 2025, p. 53 ff.

driving an autonomous vehicle, etc.; the second refers to a system that is able to adapt autonomously and solve broader tasks, regardless of the context.

But the real revolution in visual artificial intelligence came in 2022, with GenAI for speech and images. On 30 November 2022, the startup OpenAI opened a chatbot based on the Large Language Model, ChatGpt, a 'narrow' AI that can understand and generate natural language in the manner of systems such as Chatbot, Siri, Alexa or Cortana. Now starting with verbal instructions, a lot of similar type of machines responds by generating non only text, but also images, sounds, code strings or videos, becoming multimodal and establishing a dialogue with the human. Images that look like hyper-realistic or surreal photographs are beginning to spread among the general public[53]. Image production tools such as Dall-E (already active in 2021), Midjourney (2022), Stable Diffusion (2002), PointE (2022) and Copilot (2023) have begun to spread very rapidly through social media and the web in general. Images are generated based on a dataset (whose contents are known neither to the audience using it nor perhaps even entirely to the designer of the system), simulating a creative process. The system allows the image(s) given in response to be recalibrated until the image deemed 'right' by the person querying the algorithmic model is achieved.

The question then arises regarding how AI-generated images are realised, according to which ethical principles, what purpose they may serve, and what ethical implications they entail. We are moving from query logic to prompt logic. It is no longer a 'computer-assisted' image, but a 'computer-generated' image. These are images based on neural network architectures[54].

It should be noted that in order to generate images, GenAI programmes must first be able to read them and recognise their salient features: for example, in the case of a human figure, to distinguish the parts of a face, whether adult or child, or a particular

---

53 M. Boden, *Artificial Intelligence,* cit.

54 Y. Cao *et al.*, "A Comprehensive Survey of AI-Generated Content (AIGC): A History of Generative AI from GAN to ChatGPT", in «arXiv:2303.04226», March 2023.

species of animal, a certain category of objects, as well as a style, a feeling or an emotion. Machine learning systems make these processes easier and more effective. Machine learning can recognise certain configurations and link them to each other and to specific social and cultural constructs. For instance, a programme can be taught and trained to distinguish one type of animal from another, or to recognise watercolours from oil paintings or engravings. Learning can be supervised or unsupervised. In the first case, it is based on a training set chosen by the model programmers, i.e. a set of examples that will provide the target value for new images. To do this, very large databases are used, available online and organised according to specific taxonomies (ImageNet is an example)[55]. The system learns to see different things depending on its training set. In the second case, labelled examples are provided: the programme learns by analysing a huge number of images, noting the occurrence of certain associations and classifying them according to common characteristics. The accuracy and effectiveness of both depends on the quantity and quality of the input fed to the model.

AI-generated images can be created from scratch using various technologies: starting from text (Diffusion Models) or starting from an existing image (Generative Adversarial Networks, GANs). The latter types were probably the most widely used until a few years ago and offer sophisticated results in the generation of artistic artificial images. Today, Diffusion Models are widely more common to create text-driven images. Other alternatives, not as widespread as diffusion models in the field of GenAI, include VAE (Variational Autoencoders) and RNN (Recurrent Neural Networks), which employ different approaches to generate synthetic content.

Concerning GANs the invention of this typology of the visual is traditionally attributed to Ian Goodfellow and his collaborators, when they published a paper on the functioning of generative adversarial networks in 2014.[56]

---

55 On ImageNet, for example, cf. M. Coeckelberg, *AI Ethics*, MIT Press, Cambridge Mass.- London 2020, p. 129.

56 J. Goodfellow *et al.*, *Generative Adversarial Networks*,in «arXiv:1406.2661», June 2014; M. Jovanović *et al.*, "Generative Artificial Intelligence: Trends and Prospects," in «Computer», October, 2022 pp.

What distinguishes a digital image generated by 'traditional' computer graphics programs from the productions of AI relates not only to the aesthetic result, but also to the creative process. The program assumes an autonomous role in the creation and the human delegates some of the so-called 'creative' decisions.

In short, a GAN is made up of two networks that compete in a sort of 'adversarial' mode to improve by learning from each other. On the one hand, we have a *generator* network whose task is to generate new images from as large a data set as possible (in which each image is associated with a text describing its content); on the other hand, we have a *discriminator* network whose task is to compare itself with the result of the *generator* and report whether or not the image deviates too much from the human's request, whether or not it seems excessively false, whether or not it is the result of a 'hallucination' of the system, and so on. By reporting a possible error or inconsistency, the adversarial network learns and allows the other network (*generator*) to learn in turn.

As Goodefellow states:

> The generative model can be thought of as analogous to a team of counterfeiters, trying to produce fake currency and use it without detection, while the discriminative model is analogous to the police, trying to detect the counterfeit currency. Competition in this game drives both teams to improve their methods until the counterfeits are indistiguishable from the genuine articles.[57]

The final product is basically achieved when the *generator* creates a new image that is perceived by the *discriminator* as completely 'authentic'.[58]

107-112; M. Mazzone, *Le GAN e la questione della creatività nell'arte e nell'intelligenza artificiale*, in A. Barale (ed.), *Arte e intelligenza artificiale. Be my GAN*, Jaca Book, Milano 2020, pp.51-74: 51 ff; Y. Zhao, "GAN-based Image Generation," in «Applied and Computational Engineering», 92, 2024, pp. 128-135.

57 J. Goodfellow *et al.*, *Generative Adversarial Networks*, cit., p. 1.

58 The varieties of GAN, depending on the application domain, that have arisen in recent years, differ in terms of both network architecture and type of training.

Interesting is the observation of Michael Castelle[59], who interprets GANs as a form of social life. In a recovery of Pierre Bourdieu's theories, he identifies the concept of *habitus* – understood as social context – as "systems of enduring dispositions, structuring structures arranged to function as structured structures, i.e. as the principle of generating and structuring practices and representations that can be 'regulated' and 'regularised' without being at all the product of obedience to rules [...]".[60] This is a dual vision similar to that of the GAN: on the one hand, it interprets it as a descriptive structure of society as it appears, and on the other hand, as a structure that generates new practices and new ideas.

Moreover, the concept of *habitus* – a term coined by Bourdieu in 1967 when he wrote the postface to and translated Erwin Panofsky's *Gothic Architecture and Scholasticism*[61] – is theorised, starting with Thomas Aquinas who describes an initial philosophical analysis of it, as the process of acquiring the Christian virtues bestowed by the supreme good. According to Panofsky, Gothic architectural models express this disposition through an artist. Bourdieu criticises this view, re-sematising the term and refining the concept:

> [H]abitus is in no sense a mechanical principle of action or, more exactly, of reaction (it is not a 'reflex'). It is conditioned and limited spontaneity. It is that autonomous principle which means that action is not simply an immediate reaction to a brute reality, but an 'intelligent' response to an actively selected aspect of the real: linked to a history fraught with a probable future, it is the inertia, the trace of their past trajectory, which agents set against the immediate forces of the field,

---

59 M. Castelle, *La vita sociale delle reti antagoniste generative (GAN),* in A. Barale (ed.), *Arte e intelligenza artificiale*, cit., pp. 19-49.

60 P. Bourdieu, *Outline of a Theory of Practice*, Cambridge University Press, Cambridge 1977, p. 85; C.V. Dendasck and G.F. Lee, "Concept of Habitus in Pierre Bourdieu and Norbert Elias," in «Multidisciplinary Core Scientific Journal of Knowledge», 3, I, May 2016, pp. 1-10.

61 E. Panofsky, *Architecture gothique et pensée scolastique*, French transl. and postface by P. Bourdieu, « Le sens commun », Éditions de Minuit, Paris 2016; R. Alciati, *Un nuovo spirito scientifico: la rivoluzione simbolica di Pierre Bourdieu*, in P. Bordieu, *Il campo religioso*, eds. by R. Alciati and E.R. Urciuoli, Accademia University Press, Torino 2012, pp. 3-49.

> that means that their strategies cannot be deduced directly from either the immediate position or the immediate situation.[62]

Habitus is the constant tension between these two forces. It is a duality created by adaptation "between subjective structures (things, institutions) and embedded structures (dispositions, habitus)". The *habitus* can be read as a manifestation of society, made up of a static part, comparable to habits and customs, based on an almost automatic repetition of an action, and a dynamic part, which has a generative force and evolves over time. It is the coexistence of these two apparently contradictory elements that articulates the concept of *habitus*. [63]

Moreover, the *habitus* is learned through a specific form of mimesis, the imitation of practices, conscious and performative. Knowledge is thus configured for Bourdieu – inspired by the theories of Piaget and Vygotsky – as something that expands through action, through doing. Castelle notes a certain similarity between the dual *generator-discriminator* structure and the duality of the *habitus* theorised by Bourdieu. The two models that make up the GAN train themselves by refining their responsiveness as they operate through an ever-active duality, as in the *habitus*.[64]

Regarding Diffusion Models (DM) – without delving into technical definitions – these are generative models inspired by thermodynamic diffusion processes, "treating pixels like the molecules of a drop of ink spreading out in a glass of water over time."[65] They can create high-quality and extremely realistic images through textual prompts, having been trained on billions of images (datasets).

---

62 P. Bourdieu, *The Social Structures of the Economy*, Polity Press, Cambridge 2005, pp. 211-12.

63 G. Sapiro, *Habitus*, in Ead. (dir.), *Dictionnaire international Bourdieu*, CNRS Éditions, Paris 2020, p. 387 ff; R. Alciati, *Un nuovo spirito scientifico*, cit.

64 Images are generated with AI, in the field of art, including with special robots that draw on a physical medium, as Leonel Moura experimented with by inaugurating a form of robotart (L. Moura, "Robot Art: An Interview with Leonel Moura," in «Arts», 7, 3, 2018, p. 28).

65 D. Bergmann and C. Stryker, *What are Diffusion Models?* in: https://www.ibm.com/think/topics/diffusion-models.

The central idea of Diffusion Models lies in a diffusion process (forward process) that transforms an initial image into random noise. Essentially, this involves "destroying" the training dataset through a so-called noising process, where random noise with a Gaussian distribution is added over a series of time steps. Subsequently, the dataset is reconstructed step by step by reversing the initial process (the so-called denoising or reverse process), generating a new, high-quality, and plausible image. During this phase, the model generates a sequence of intermediate images that progressively become less noisy until reaching a detailed and recognizable final image consistent with the prompt's requirements.

These Diffusion Models are trained on a large set of images. During training, the model learns to predict the noise added to an image at each timestep of the diffusion process so it can then subtract it correctly in the reverse process. The randomness of the noise also explains why diffusion models generate different images each time, even when the same prompts are used.

The prompt (as in the case of DALL-E or Stable Diffusion, which are conditional diffusion models) guides the generation process, ensuring that the final image aligns with the user's request. This conditioning can be implemented using techniques such as text embedding, where the text is transformed into a numerical representation that guides the reverse diffusion process.

The training process can use a function (such as the mean squared error, MSE) to measure the difference between the noise estimated by the model and the actually added noise. This function helps the model identify and remove the noise with greater accuracy.

Additionally, a classifier within the model can assess the quality and coherence of the generated image with respect to the initial prompt. This classifier facilitates alignment and consistency between the produced image and the prompt request[66].

---

66 V. Bok, *Image Generation Models. GANs, Diffusion Models, and Transformers*, Manning Publications, New York 2025; O. Sanseviero *et al.*, *Hands-On Generative AI with Transformers and Diffusion Models*, O'Reilly Media, Sebastopol 2024.

What emerges from the foregoing is how artificial images provoke reflections on multiple spheres, from the creative and aesthetic, to the technological, social and ethical. On this last aspect we can observe that we add the creation of another imaginary that could be called 'transgenerational', as it facilitates dialogue between multiple generations, albeit not necessarily in a direct way; [67] at the same time, it seems to be generated by the iconographic material found on the web, an ever-expanding dataset that is partly shaped by chance, based on the input of the online public, partly by the policies of large corporations, and again in relation to the orientations of the so-called cancel culture of the moment.

As we shall see, the imaginaries, prejudices and stereotypes we are talking about are those that reflect the mainstream cultures of the last thirty years, the period of the massive spread of the Internet and smartphones, with which photos can be taken and shared quickly, and now with which they can be created with AI.

The use of AI in targeted disinformation campaigns can result in deceptively realistic and potentially powerful outcomes. It reflects a plurality of possibilities of the real, rather than representations of something existing or like the existing.

It is not uncommon for artificial images to be perceived as hallucinations, the product of altered states of consciousness, or as erroneous or deceptive outputs generated by AI models.[68]

These images are generated without reproducing physical reality, but rather by simulating it, with the precise nature of the reality being simulated remaining unknown. Such systems make use of processes whose workings we cannot know. Chatonsky, not surprisingly, talks about "hallucinating images".[69] These images are certainly not made *by* a human being, but neither are they consciously produced *for* a human being. As the artist Trevor Paglen writes, "[v]isual culture

---

67 See also about the concept of 'transgenerational: T. Andina, *Digital Ethics: The Issue of Images*, Bloomsbury Publishing, London 2022, p. 160 ff.

68 G. Finocchiaro, *Intelligenza artificiale. Quali regole*, Il Mulino, Bologna 2024, p. 24.

69 G. Chatonsky, "Deep Dream (The Network's Dream)," in «SubStance», XLV, 2, 2016, pp. 61-77: 63.

has changed form. It has become detached from the human eye and largely invisible. Human visual culture has become a special case of seeing, an exception to the rule. The overwhelming majority of images are now made by machines for other machines, with humans rarely in the loop",[70] that is they are made by a machine and not necessarily for a human audience.

These are programs that require a quantity of training images that no human mind could contain in its memory because of all its visual experiences. The visual universe of these programs is beyond our reach and does not adhere to the more general afferents and relations of the human mind. Therefore, we can only try to get to know this visual universe by practising with these tools and analysing their inputs and outputs. The processes of generation are alien to us.[71]

As Mlađan Jovanović writes, "[g]enerative artificial intelligence can make powerful artifacts when used at scale, but developing trust in these artifacts and controlling their creation are essential for user adoption".[72]

---

70 T. Paglen, "Invisible Images (Your Pictures are Looking at You)," in «The New Inquiry», 2016: https://thenewinquiry.com/invisible-images-your-pictures-are-looking-at-you/; K. Crawford and T. Paglen, "Excavating AI: the Politics of Images in Machine Learning Training Sets," in «AI & Society», 36, 4, 2021, p. 1399; F. Restuccia, *Immaginazione artificiale? Flusser e l'intelligenza artificiale generativa*, in A. Fabris (ed.), *Immagini, media, dispositivi. Studi e prospettive*, Edizioni ETS, Pisa 2023, p. 72.

71 Artists, advertisers, journalists and the public are approaching image-generating artificial intelligence experiments. Consider the video series *Hyperdimensional Attractions* (2019) by Mario Klingemann, in which figures gradually mutate into others in a seamless metamorphosis of images. The images may be based on data sets or sensors that pick up what is happening in the environment. As in other works by the same artist, the installation *Memories of Passersby I* (2018) or the work *Uncanny Mirror* (2018), the same principle is taken up: the camera captures the viewers and distorts what they see in a disturbing way. Cf. M. Klingemann, *Memories of Passersby I*, in A. Barale (ed.), *Arte e intelligenza artificiale*, cit., pp. 75-88 and Id., *Uncanny Mirror*, in A. Barale (ed.), *Arte e intelligenza artificiale*, cit., p. 97.

72 M. Jovanović and M. Campbell, "Generative Artificial Intelligence: Trends and Prospects," in «Computer», October 2022, p. 107.

We can observe that image-generative AI introduces complex and specific ethical implications, which will be discussed in the following chapters. For instance, the absence of emotion in an image-generating engine may influence the choices made by individuals who request and subsequently use these images.

In fact, Nello Cristianini states that “As we move away from an anthropocentric view of intelligence, we should not expect sentience or consciousness in our machines, the spontaneous emergence of a language that is translatable into ours, and emotional traits such as envy or fear that might afect our relation to these agents. All these are just projections of our incapability to imagine truly alien intelligences”.[73]

With the spread of AI, in the “fourth revolution” *à la* Floridi, we have probably lost our central position in the world and in our own minds.

The *algorithmic turn* has significantly impacted the visual domain. Image disciplines now intersect with extra-media and extra-aesthetic areas, including data science, surveillance processes, scientific research, and uses in information, advertising, public opinion, and art.

Ethical issues concern not only the values embedded in the models but also their applications. These are distinct yet complementary concerns, both equally important.

Reflecting on the ethical implications of AI, we can draw parallels to the myth of Prometheus and the significance of fire.

> And Prometheus, not knowing how he could devise man’s preservation, stole the wisdom of practicing the arts of Hephaestus and Athene, and fire with it [...], and gave them to man. [...] And in this way man was well supplied with the means of life. But Prometheus is said to have been afterward prosecuted for theft, owing to the blunder of Epimetheus.
>
> Now man, having a share in divinity, was at first the only one of the animals who had any gods, because he alone was of their kindred, and he would raise altars and images of them. He was not long in inventing

73 N. Cristianini, *The Shortcut. Why Intelligent Machines Do Not Think Like Us*, Routledge, London 2023, p. 13.

> articulate speech and names; and he also constructed houses and clothes and shoes and beds, and drew sustenance from the earth.[74]

The hero stole fire from the gods to bring knowledge and freedom to mankind. AI, like fire, can offer great opportunities, but also risks in terms of how it can be used for good or evil.

The possibility of establishing a trusting relationship with image-generating systems can be based on good literacy, an awareness of the potential and limitations of such AI tools, starting with the new imaginative paradigm that is being created, and not secondarily with the rules and standards that States are slowly equipping themselves with.[75]

1.4. *Artificial Intelligence and the Facets of Imagination*

In the light of what has been said about algorithmic imaging, it seems appropriate to wonder whether, from an ethical standpoint, image generators based on AI represent a comparable radical breakthrough to those witnessed in the past, with the advent of photography in the nineteenth century and the emergence of digital images and the Internet in the late twenty-first century, when compared to traditional engravings and images in general.

It therefore seems essential, firstly, to ask whether (and in what way) the individual's approach to and action through the visual is changed by such tools. Secondly, in order to really understand the first aspect, it seems necessary to explore whether and to what extent the individual's way of imagining and understanding the visual has changed, and with what implications for our actions, both individual and collective, for our choices and decisions – as we shall see in the following chapters. These two aspects seem to be closely intertwined: just as the image-generating AI changes our relationship to the

---

74 Plato, *Protagoras*, Engl. transl. by B. Jowett and M. Ostwald, The Library of Liberal Arts, Bobbs-Merrill, Indianapolis – New York 1956, 321c-322a, pp. 19-20.

75 The reference can only be to the AI Act (Regulation (EU) 2024/1689), to which we will turn further on in Chapter 3. See https://artificialintelligenceact.eu/.

visual, it also changes our way of imagining, which in turn changes again our way of engaging with the visual, creating a seamless cycle.

Regarding to the first aspect, our approach to the visual, we have been enveloped in a relentless web of digital images for more than three decades, since the advent of the Internet and its widespread use. With social media and smartphones, this process has intensified. These images are constantly moving and potentially changing, while maintaining a lowest common denominator, the shape of the rectangle. The rectangle represents a window into another 'world', a passage from one dimension to another. This was first stated in a revolutionary way by Leon Battista Alberti in the fifteenth century in *De Pictura* (1435) and still holds true today in the universe of digital and AI. If the painting can metaphorically represent a window, its frame defines its figurative space and holds the observer's gaze by imposing his attention on what it delimits. Falcinelli argues that the dominance of the rectangle from the Renaissance onwards was linked to four specific events: the birth of perspective, the spread of painting on canvas even to small bourgeois patrons, the invention of printing (and the consequent arrangement of rows, columns and drawings within a rectangle) and Alberti's theory of the rectangle as a window on the world. The rectangle became, to all intents and purposes, a social object.[76]

If the form of access to the visual seems to have remained unchanged from the sixteenth century to the dawn of AI – computer screens, tablets, smartphones, etc. –, the way we interface with the visual is changing. We increasingly get information, communicate and act with and in relation to the visual. We use intermediary

76 L.B. Alberti, *De Pictura* (1435), edited by L. Bartolini, Polistampa 2012 [*On painting*, transl. with introduction and notes by J. R. Spencer, Yale University Press, New Haven – London, 1966]; R. Falcinelli, *Figure. Come funzionano le immagini dal Rinascimento a Instagram*, Torino, Einaudi 2020, pp. 71, 74-75, in which he also writes: "The computer screen, the TV screen, the smartphone screen are rectangles, the cartoons of comic strips, and the paintings of video games, almost all paintings and all photographs are rectangles. News reports and football matches are seen within a rectangle, and so is a dentist's X-ray, a map of a city and even a pornographic video. Even the book you are reading is a rectangle" (p. 74).

elements that enable a sensitive perception of what is in front of our eyes, a concept similar to the *diáphanes* mentioned in Aristotle's *De Anima*, diaphanous and transparent elements through which we see clearly, making the image of an object recognisable, distinguishable to the eye.[77] With AI, we are encountering both hyper-real worlds and, at the same time, hyper-fantastical worlds in immersive modes. Consider the worlds created by virtual reality (VR) and augmented reality (AR), immersive experiences made possible by the various emerging forms of reality.[78] We can simulate living in artificially constructed dimensions – consider the *Second Life* project and the more recent Metaverse – to the point of believing that we actually live there.[79] We therefore live in a hybrid dimension, "onlife",[80] where the image has become the preferred means of communication. As Chalmers writes:

---

77 Cf. Aristotle, *De Anima*, II.7 418b4-9, in *The Works of Aristotle: De Anima*, Engl. transl. by J.A. Smith, Clarendon Press, Oxford 1931 : "[B]y 'transparent' I mean what is visible, and yet not visible in itself, but rather owing its visibility to the colour of something else; of this character are air, water, and many solid bodies. Neither air nor water is transparent because it is air or water; they are transparent because each of them has contained in it a certain substance which is the same in both and is also found in the eternal body which constitutes the uppermost shell of the physical Cosmos".

78 D.J. Chalmers, *Reality+: Virtual Worlds and the Problems of Philosophy*, cit.

79 If AR (augmented reality) is an object reality consisting of a physical space to which multimedia elements are added, allowing individuals to live an interactive experience and augmenting the human experience with mobile technological devices that use AI systems to interact with reality, with virtual reality, on the other hand, defines a new world that does not exist in any other dimension of object reality. See M. Bergamasco *et al.*, *The Impact of Virtual and Augmented Reality on Individuals and Society*, Frontiers Media SA, Lousanne 2019; R. Doerner and W. Broll (eds.), *Virtual and Augmented Reality (VR/AR): Foundations and Methods of Extended Realities (XR)*, Springer Nature, Switzerland AG 2022. On Second Life and Metaverse: P. Canastrari and A. Romeo (eds.), *Second life. Oltre la realtà il virtuale*, Lupetti, Milano 2008; M. Ball, *The Metaverse: And How It Will Revolutionize Everything*, WW Norton & Co, New York 2022; D.J. Chalmers, *Reality+*, cit., pp. 265 ff.

80 L. Floridi, *The Onlife Manifesto: Being Human in a Hyperconnected Era*, Springer, Cham 2014.

> VR is not illusory [...] VR is artificial, but so are many non-virtual environments. Many of us live in cities that are largely human-made, but we still manage to lead meaningful and valuable lives. So, the artificiality of an environment is no bar to value. It's true that some people value a natural environment, but this seems to be an optional preference: it's equally possible to prefer an artificial environment, and there's nothing irrational in doing so. Even for people who prefer a natural environment, life in an artificial environment will often be a life worth living.[81]

If anything, it seems that the eye has become the organ of choice for operating in realities and guiding our choices. As Derrick De Kerckove[82] points out, technological innovations bring about changes in the habits and attitudes of individuals. New perceptual and mental models of the world are constructed. Scientists such as Wolf and Small claim that digital technology is changing our brains at the cellular level. New neural pathways are being strengthened.[83] Belting himself observes that the dialogue between increasingly artificial endogenous and exogenous images takes place in our brains.[84] And the speed with which they arrive in the mind of the viewer runs the risk of transcending their deeper meaning or the information they convey in favour of their spectacularisation.

What we want to consider here, then, is whether and to what extent our way of imagining, of 'seeing' with our minds, and thus of producing images, changes with the advent of images generated by AI. A change in the way we imagine, if radical, may produce different modes of action and new imaginaries within which to make choices.

---

81 D.J. Chalmers, *Reality+*, cit., pp. 306-307.

82 He introduces the concept of "technopsychology" concerning the conditioning that technologies take on our minds and our reality. See D. De Kerckhove, *The Skin of Culture: Investigating the New Electronic Reality*, Kogan Page, London 1998.

83 M. Wolf, *Reader. Come Home: The Reading Brain in a Digital World*, Harper Collins, New York 2018.

84 H. Belting, *An Anthropology of Images: Picture, Medium, Body*, Princeton University Press, Princeton & Oxford 2011.

Can GAN's or diffusion models' answers to our questions be seen as the result of a form of creativity[85] and imagination? Or is imagination the new visual product? Some scholars claim that there may be a form of AI imagination with which human imagination can converse, or at least interface.[86] Is it a question of both human and machine imagination? Or is it about delegating our imagination to machines? Are these paradigm shifts in the visual changing the ethical horizon of our actions?

First of all, it is necessary to begin with the transversal and multifaceted concept of 'human' imagination and then move on to the contemporary variations of technical, artificial, extended, augmented, mixed and collaborative imagination. Throughout the history of philosophical thought, this concept has taken on a variety of meanings and nuances. We will summarise just a few of them here, which are functional to understanding the role that imagination plays today in relation to the spread of AI and the impact that these technological changes are having on the way we imagine and the preferences of individuals.[87]

By imagination, from the Greek *eikasìa* or *phantasìa* and the Latin *imaginatio*, we can mean, according to the Oxford Dictionary definition, "the ability to create pictures in your mind; the part of your mind that does this or that you have imagined rather than something

85 On creativity, Boden states that there are "three types of creativity": combinatorial, exploratory, and transformative. It is "the ability to come up with ideas that are new, surprising, and valuable". Cf. M.A. Boden, *Creativity and Art: Three Roads to Surprise*, Oxford University Press, Oxford/New York 2010, p. 1; Ead., *The Creative Mind: Myths and Mechanisms*, Routledge, London 2004, pp. 1-10; Ead., *Creativity and Artificial Intelligence: A Contradiction in Terms*?, in E.S. Paul, S.B. Kaufman (eds.), *The Philosophy of Creativity: New Essays*, Oxford University Press, Oxford/New York 2014, pp. 224-44.

86 E. Finn, *What Algorithms Want. Imagination in the Age of Computing*, MIT Press, Cambridge Mass 2017.

87 For an in-depth examination, see, A. Ferrarin, *Un mondo non di questo mondo*, cit.; M. Ferraris, *L'immaginazione*, cit., J.J. Wunenburger, *L'imaginaire*, PUF, Paris 2020; Id., *Le vie des images*, cit., V. Neri, *L'immagine nel web. Etica e ontologia*, Carocci, Roma 2013, pp. 17 ff.

that exists or the ability to have new and exciting ideas".[88] According to the Treccani Italian dictionary, we can broaden the range of definitions by characterising imagination as "a particular form of thinking that does not follow fixed rules or logical connections, but presents itself as a free reproduction and elaboration of the content of a sensory experience, linked to a specific affective state and often oriented towards a fixed theme; it can give rise to dream-like activity (as in so-called 'daydreams') or harmonious creations with artistic content (artistic imagination), or even, through a mechanism related to intuition, to conclusions rich in practical content; in a more general definition, it is the ability to form, elaborate, develop and even deform images, in each case as a creative force."[89]

The term imagination therefore seems to be related to the concept of fantasy, the creative faculty of "preserving and reproducing within oneself sensible perceptions, thus preparing the memory".[90] As Munari writes: "Imagination / Everything that was not there before, but unrealisable / Invention / Everything that was not there before, but exclusively practical and without aesthetic problems / Creativity / Everything that was not there before, but realisable in an essential and global way. Imagination / Invention, creativity think, imagination sees".[91] Fantasy would then seem to be the 'visual' and most radical heart of imagination.

According to Ferraris, "imagination is the preservation of the absent, fantasy its reworking", so it is more akin to the realm of the

---

88 Headword *Imagination*, in Oxford Dictionary online: https://www.oxfordlearnersdictionaries.com/definition/english/imagination.

89 Headword *Immaginazione* in Vocabolario Treccani online: https://www.treccani.it/vocabolario/immaginazione/.

90 The affinity between imagination and fantasy seems to be gradually weakening at the end of the modern age. On the lexical question of the concepts of imagination and fantasy and their relationship, see in particular M. Ferraris, *L'immaginazione*, cit., pp. 7-13; the headword *Immaginazione*, in G. Devoto and G.C. Oli, *Dizionario della lingua italiana*, Le Monnier, Firenze 1995, p. 925; the headword *phantasía*, in L. Rocci, *Vocabolario Greco-Italiano*, Società Dante Alighieri, Roma 1988, p. 1941.

91 B. Munari, *Da cosa nasce cosa*, Laterza, Roma-Bari 1981; Id., *Fantasia*, Laterza, Bari-Roma 1977, pp. 8-15.

unreal. It can belong to the mental image, to the sphere of invention, or to belief/illusion, according to Strawson's famous tripartition.[92]

The object of imagination or fantasy is called an *eídolon*, idol or *phantásma*. The concepts of imagination and fantasy both refer to the ability to mentally retain the product of sensible perceptions. Over time, the meaning of the two terms, which in many ways coincide, has been distinguished in that imagination has been assigned the role of generating mental reproductions of objects of reality in ever new associations, in obedience to practical interests, or of creating images, while fantasy has been assigned the role of aesthetic elaboration of mental representations or images. Imagination or fantasy produces not only visual mental images, but also tactile, acoustic, olfactory and motor images. In the creative act, these mental images can be expressed in works of various kinds, such as paintings, sculptures, poems, music, dance, etc.

The first attestation of the term phantasy, translated as imagination, is attributed to Aristotle, who traces its etymology back to the Greek word *phantasìa*[93] and thus to the root *pháos*, which can be linked to the verb *phántomazomai*. Aristotle's *phantasìa* already refers to the ability to produce sensible images (*phantásmata*), a form of *kínesis* that occurs in sentient beings and enables them to represent images both on the basis of objects that can be perceived by the senses and *in absentia*. He defines it as a function of primary importance for the acquisition of intellectual knowledge, to be distinguished from sensation and opinion.[94] It assumes an intermediate, mediating function, as Flusser would later write, leading from sensation to thought and the intellectual faculty. A concept later taken up in the early sixteenth century by Gianfrancesco Pico della Mirandola in *De*

92 M. Ferraris, *L'immaginazione*, cit., p. 7; P.F. Strawson, *Freedom and Resentment and Other Essays* (1974), Routledge, London 2008, pp. 50-72.

93 Aristotle, *De Anima*, cit., III.3 429a 2; Id., *De memoria et reminiscentia. Memoria e reminiscenza*, edited by G. Nastasi, Ist. Italiano Studi Filosofici, Napoli 2024. Cf. also A. Astolfi, *Phantasia in Aristotele*, Vita & Pensiero, Roma 2011; M. Nussbaum and A.M. Rorty (eds.), *Essays on Aristotle's De anima*, Clarendon Press, Oxford 1992.

94 Headword *Immaginazione*, in «Treccani Filosofia», cit., vol. I, p. 500.

imaginatione, which identifies *imaginatio* with the "motus is animae quem sensus, in actu positus, parit". It is an imagination that can become erratic. Only reason, and then faith, can lead it to a more authentic and pure knowledge.[95]

For Plato, too, imagination (*eikasía*) is part of the cognitive process, linked to the sensory image of an object received externally or through divine action, and then becoming an autonomous creative activity of which man alone is capable, performing a cognitive mediation between the senses and the intellect. In Book IV of the *Republic*, *eikasía* is described as a vision of shadows and appearances reflected on smooth, shiny surfaces. By imagination, then, Plato means both true representation and illusion. [96]

The Aristotelian and Platonic views remained valid in the Western world until Plotinus, who recovered the two theories and partly merged them.[97] For him, the imagination produces visual mental representations. At the risk of oversimplifying, it can be said that he offers two interpretations of the imagination – although Alessandro Linguiti states that it is not possible to speak of a dualistic theory *tout court* – one with a more sensory value, the other, intellectual, according to which the imaginative faculty comes from the intelligible, which becomes conscious thanks to the imagination, in

95 G. Pico della Mirandola, *L'immaginazione*, edited by F. Molinarolo, Edizioni della Normale, Pisa 2022, pp. 119-21.

96 Plato, *Timeo*, in Id., *Timaeus; and, Critias*, ed. by D. Lee, Penguin, London 1977, 71a, p. 262 ff.; Id., *Republic*, ed. by R. Waterfield, Oxford University Press, Oxford 2008, 509d-510; Aristotle, *On Sleep and Waking*, III, 461a, 2 ff. (in *On Sleep and Dreams*, Engl. transl. by D. Gallop, Liverpool University Press, Liverpool 1996), but also *Theaetetetus* and *Philebus*. See also L.M. Napolitano Valditara, *Platone e le "ragioni" dell'immagine: percorsi filosofici e deviazioni tra metafore e miti*, Vita & Pensiero, Milano 2007, p. 238.

97 In the Roman world, Greek *phantasía* is present in three lexical variants, listed in chronological order of appearance: *visio*, already in Cicero, *imaginatio*, a later term, but also present in Pliny the Elder and, with a more realistic flavour, the transliteration of Greek *phantasía* (in the sense of extravagance). On the concept of fantasy and imagination in the ancient world, see M. Ferraris, *L'immaginazione*, cit., pp. 7-12.

order to give form to the formless.[98] The latter theory was still valid at the height of the Renaissance with Marsilio Ficino and Giordano Bruno, who linked imagination to the art of memory, the ability to form images (and to retain the idea and meaning of them). Today, such an approach would be linked to a neurobiological theory of imagination.[99]

During the scientific revolution, the debate about the imagination became intense again: the term 'fantasy' – until then almost synonymous with imagination – was confined to the realm of unreality.

In the course of the seventeenth century, within the framework of English empiricism, the imagination represented the ability to make connections between even very distant elements, without any cognitive power. A debate also developed about the imagination's ability to shape objective reality and the ethical consequences that might follow from this. The rationalist current, led by Descartes and Nicolas Malebranche, devalued the imagination from a gnoseological perspective, nevertheless, recognised a "power of sensibility".[100]

---

98 Plotinus, *The Enneads*, ed. by R. Waterfield, Oxford University Press, Oxford 2008, IV, 4, 13. Cf. A. Linguiti, *Immagine e concetto in Aristotele e Plotino,* in Cl.-C. Härle (ed.), *Ai limiti dell'immagine*, Quodlibet, Macerata 2005, pp. 69-80, in part. pp. 73-77.

99 G. Bruno *Sigillus sigillorum* (1583), in Id., *Opera latine conscripta*, II, 2, edited by F. Fiorentino, F. Tocco, H. Vitelli, *et al.*, Neapoli Florentine 1879-1891, p. 208. On neurobiology of imagination, cf. L.F. Agnati *et al.*, "The Neurobiology of Imagination: Possible Role of Interaction-dominant Dynamics and Default Mode Network," in «Frontiers in Psychology», 24, 4, 296, 2013; M. Denis *et al.* (eds.), *Imagery, Language and Visuo-Spatial Thinking*, Psychology Press, London 2012.

100 Malebranche defined imagination as "une folle qui se plaît à faire la folle". Cf. N. Malebranche, *De la recherche de la vérité* (1674), in Id., *Œuvres complètes*, edited by G. Rodis-Lewis, vol. II, Vrin, Paris 1962, p. 261. In *Meditationes de prima philosophia* (1641), Descartes explicitly distinguishes between the imagination and the intellect. Moreover, he associates it with memory, as did Ficino and Bruno. An imagination that is in opposition to reason and *bon sens*. On Descartes, in particular, cf. *Discours de la méthode* (1637), *Méditationes de prima philosophia* (1641), *Passions de l'âme* (1649), in Id., *Œuvres complètes*, eds. M. Beyssade and D. Kambouchner, Gallimard, Paris 2018. Cf. M. Ferraris, *L'immaginazione*, cit., pp. 67-79, 75 ff.; E. Cassirer, *The Philosophy of*

Similarly, Spinoza defines the imagination as an opinion, a source of illusion, and therefore deceptive and opposed to reason. In contrast, Thomas Hobbes, in *Leviathan* (1651), understands the imagination as an attenuated sense that offers sensible images and is connected to sensation.[101] Leibniz, in his *Nouveaux Essais sur l'entendement humain* (1765), perceives the power of the imagination, whereas John Locke (1690) considers the imagination to be prior to experience. Locke posits that the imagination arises from the senses and gives rise to ideas.[102]

In the Enlightenment, the concept of imagination and of fantasy are contrasted with that of *realitas*. In the entry for d'Alembert and Diderot's *Encyclopédie*, Voltaire offers an intriguing perspective on the nature of imagination: "Imagination is the power that every sentient being feels within themselves to represent sensible things in their mind; it is a faculty that depends on memory. We see people, animals, gardens; these perceptions penetrate us through the senses, the memory preserves them, the imagination composes them; that is why the ancient Greeks called the Muses 'daughters of memory'."[103]

In the nineteenth century, on the other hand, we witness an idea of imagination that breaks up and reassembles itself in new forms, as in the concept of the imaginary. The dialectic between imagination

---

the Enlightenment, Engl. transl. by C.A. Koelln and J.P. Pettgrove, Princeton University Press, Princeton 1951, p. 353 ff.

101 Chapter One, but especially in Chapters Two and Three of *Leviathan, or the Matter, Form and Power of an Ecclesiastical and Civil Commonwealth* (1651), edited by C. Brooke, Penguin Classics, London 2017.

102 J. Locke, *An Essay Concerning Human Understanding* (1689), edited by K.P. Wrinkler, Hackett Publishing, Indianapolis 1996, pp. 386 ff. and G.W. Leibniz, *New Essays on the Human Intellect* (1765), It. transl. edited by D.O. Bianca, *Nuovi saggi sull'intelletto umano*, Utet, Torino 1976, in particular p. 168 and pp. 650-52; D. Hume, *A Treatise of Human Nature* (1738), ed. by L.A. Selby Bigge, Clarendon Press, Oxford 1896.

103 Voltaire, *Imagination*, in D. Diderot, J. D'Alambert, *L'Encyclopédie*, 1re éd., David l'Aîné Briasson, Durand Le Breton, Paris 1766, Tome 8, pp. 560-564; Id., *Immaginazione*, in D. Diderot *et al.*, *L'estetica dell'Encyclopédie*, edited by M. Modica, Editori Riuniti, Roma 1995. For a critical commentary on Voltaire's entry, see R. Vecchia, headword *Immaginazione*, in R. Romano (ed.), *Enciclopedia Einaudi*, vol. VII, Einaudi, Torino 1979, pp. 43-48.

as a voluntary, reproductive act and fantasy as an involuntary, arbitrary, creative act reaches an even more radical extreme.[104] For Schelling, in line with other Romantic positions, imagination or fantasy represents a productive faculty that provides access to the Absolute. Imagination has the task of producing 'unification', that is, the interpenetration of the universal and the particular, in which the finite and the infinite are one and the same.[105]

The concept of imagination thus acquires a much more articulated value, and the origin of this change can be traced back to Immanuel Kant at the end of the eighteenth century. In the *Critique of Pure Reason*, he distinguishes the imagination between the reproductive imagination (*reproduktive Einbildungskraft*), the faculty that allows us to represent an object in ourselves, through our sensibility, and productive imagination, the "*a priori* faculty of synthesis" (*produktive Einbildungskraft*), a source of knowledge. Kant thus seems here to take up the Aristotelian theory of the cognitive value of imagination. The imaginative controls itself and sets limits to the unreal, while fantasy is ascribable solely to the sphere of the irrational. Imagination has also a mediating function: in the *Critique of Judgement* and in the second edition of the *Critique of Pure Reason*, imagination assumes an intermediary role between intellect and sensibility, which differ in their transcendental function of knowledge construction.[106]

---

104 In German Romanticism, imagination came to be identified with the cognitive faculty and artistic imagination, or aesthetic imagination, with intellectual intuition. In the same years, the poet Novalis emphasised the role of the productive activity of the imagination. C. Taylor, *Sources of the Self: The Making of the Modern Identity*, Harvard University Press, Cambridge Mass. 1989, pp. 427 ff.

105 In some passages of his work, Schelling actually distinguishes fantasy (*Phantasie*), the intuitive-creative faculty, from imagination (*Einbildungskraft*), the power that realises what the imagination intuits. Cf. R. Diodato, *L'arte come categoria estetica. Un'introduzione*, Eupress FTL, Lugano 2005, p. 93.

106 Cf. E. Franzini, *Un'immaginazione per il Settecento*, in F. Carmagnola and V. Matera (eds.), *Genealogie dell'immaginario*, UTET, Torino 2008, pp. 17-18, where the ambiguous role of imagination in Kant's schematism is also highlighted. Cf. I. Kant, *Kritik der Reinen Vernunft*, in „*Kant's*

For hist part, Hegel, like Kant, distinguishes between the imagination (*Einbildungskraft*) understood as a 'merely passive' reproductive faculty, of images, which therefore disintegrates and dissolves, and fantasy (*Bildungskraft* or *Phantasie*, or creative imagination), with a creative function, a memorising and revealing power of the cohesion of absolute reason. He bases the concept of genius on the latter. Imagination grasps "the reality of things under its diverse forms, an attention which, without cessation, watches everything that strikes the eyes and the ears, engraves upon the memory the images of things, and preserves them in all their variety and unchanging accuracy."[107]

In the second half of the nineteenth century, the imagination is rather attributed the function of historical reconstruction of human consciousness through its artistic productions and historical manifestations. For Marx, it seems to be the tool for disrupting action and managing its consequences, while for Gustave Le Bon (1895), imagination guides crowds and can distort the vision of an event. Georg Simmel, for his part, sees the imagination as an amplification of emotional feeling and reinforces the cognitive power of the image.[108] Weber, too, emphasises its cognitive character, which influences social action to the point of arriving at the slogan taken up by Bronislaw Baczko, "All power to the imagination",

---

Gesammelte Schriften, hrsg. Von der Königlich preußischen Akademie der Wissenschaften", Band III, Berlin 1904, Engl. transl. And edited by P. Guyer and A. Wood, *Critique of Pure Reason* 1781/1787, Cambridge University Press, Cambridge 1997, § 24 ; I. Kant, *Kritik der Urteilskraft* (1790), in „*Kant's Gesammelte Schriften*", hrsg. Von der Königlich preußischen Akademie der Wissenschaften, Band V., Berlin 1908, ed. and transl. by P. Guyer, transl. E. Matthews, *Critique of the Power of Judgment*, Cambridge University Press, Cambridge 2000. Cf. J.-J. Wunenburger, "L'image, reproduction, transformation, création de l' "irréel" ? Quelques notes sur l'anthropologie de l'imagination," in «IRIS», 44, 2024: https://publications-prairial.fr/iris/index.php?id=3615; M. Feyles, "Kant e la fenomenologia dell'immaginazione", in «Aesthetica Preprint», 120, 2022, pp. 65-78.

107 G.S. Morris (ed.), *Hegel's Aesthetics*, S.C. Criggs & Company, Chicago 1885, p. 99 ff. See also G.W.F. Hegel, *Aesthetics* (1835), by M. Knnx, Clarendon Press, Oxford 1975, pp. 280 ff.

108 G. Le Bon, *Psychology of Crowds* (1895), Sparkling Books Edition 2022.

an imagination that wants to make itself real in society, to become public, as Cornelius Castoriadis would argue.[109]

What happened with Jean-Paul Sartre in the twentieth century is interesting in terms of its role in AI. Following in the footsteps of Husserl, who considered the imagination to be endowed with an intentionality capable of grasping the essence of things, Sartre sought to transcend it, first in *L'imagination* (1936) and four years later in *L'imaginaire* (1940).[110] Sartre remains anchored to the more traditional idea that the imagination is the "fonction 'irréalisante' de la conscience". Taking up Husserl again, Sartre examines the differences between perception and imagination, which are regarded as two functions of consciousness that never operate simultaneously.[111] Imaginative activity produces unreal or absent objects, whereas perceptual activity produces real and present objects. Imaginative consciousness is interpreted as a more primordial aspect than perception. Imagination does not produce real data but shows unreal objects and frees consciousness to express itself.[112] In *L'imaginaire*, Sartre writes that the act of imagining is constitutive and at the same time isolating and annihilating ("ainsi l'acte imaginatif est à la fois, constituant, isolant, et anéantissant"), different from reality: by denying it, it turns it into an unreal object. Imaginative consciousness thus has a nullifying power, and the imaginary reduces the real to nothingness.[113] The ethical role is also

109 B. Baczko, *Les imaginaires sociaux. Mémoirs et espoirs collectifs*, Payot, Paris 1984, p. 12; C. Castoriadis, *L'institution imaginaire de la société*, Éditions du Seuil, Paris 1975, Engl. transl. by K. Blarney, *The Imaginary Institution of Society*, Polity Press, London 1987.

110 J.-P. Sartre, *L'imagination* (1936), PUF, Paris 1989; Id., *L'Imaginaire. Psychologie phénoménologique de l'imagination*, Gallimard, Paris 1986.

111 Husserl underlines how the imagination represents lived experiences in the form of fantasies. Cf. E. Husserl, *Ideen zu einer reinen Phänomenologie und phänomenologischen Philosophie*, Niemeyler, Halle 1913, Engl. transl. by F. Kersten, R. Rojcewicz and A. Schuwer, *Ideas Pertaining to a Pure Phenomenology and to a Phenomenological Philosophy*, Kluwer Academic Publishers, Dordrecht 1982.

112 J.-P. Sartre, *L'imagination*, cit., p. 32.

113 J.-P. Sartre, *L'imaginaire*, cit., p. 348 and 351: « Nous saisissons à présent la condition essentielle pour qu'une conscience puisse imager : il faut qu'elle ait la possibilité de poser une thèse d'irréalité ».

implied, not surprisingly, in the words of Wunenburger, according to whom "imagination indeed compels us to formulate an ethics, even a wisdom of images".[114]

More specifically, as regards the line taken by the French school at the turn of the Second World War, its connection with anthropology and psychoanalysis can be attributed to the (relatively) recent discovery of unconscious spaces and the importance of symbols in the communicative context. We can at least mention Gaston Bachelard, Gilbert Durand and Paul Ricœur. For Bachelard, imagination is a function of the unconscious of the dreaming individual, and the dream products that emerge from it connect him to the cosmos. It is certainly a creative imagination and represents the individual's first form of perception of the world. A theory far removed from the 'imagination' of algorithms. Durand, for his part, is concerned with the symbolic imagination, which unites consciousness with the reality of objects and assumes the function of psychosocial balance. The imagination's own dynamism is the factor of homogeneity in representation.[115] In the five lectures given between 1973 and 1974 at the Centre de Recherches Phénoménologiques in Paris, Ricœur addresses the polysemy of the senses of the imagination. He asserts that the imagination is the resonance in us of a new being of language. It reawakens our sensory and emotional faculties through the dynamic interplay and the semantic potential that language can reveal.[116] Ricœur theorises a creative imagination that is closely

---

114 Cf. J.-J. Wunenburger, *L'imaginaire*, cit., p. 117.

115 The symbolic imagination assumes four functions according to Durand: the biological function; the psycho-social function; the humanist function; and, finally, the theophanic function, through which the symbolic imagination leads towards an infinite transcendence that stands as the supreme value. Cf. G. Durand, *L'imagination symbolique* (1964), PUF, Paris 1984, pp. 117 ff.

116 P. Ricoeur, "Les directions de la recherche philosophique sur l'imagination", in «Recherches phénoménologiques sur l'imaginaire I», Centre de Recherches Phénoménologiques, 29, Paris, 1974, pp. 1-8. The theme of imagination also recurs in other writings by Ricœur: Id., "The Metaphorical Process as Cognition Imagination and Feeling," in «Critical Inquiry», 5, 1, 1978, pp. 143-59; Id., "La Bible et l'imagination," in «Révue d'Histoire et de Philosophie religieuses», 62, 1982.

linked to the 'living' metaphor, which addresses art and opens up to the social through the concepts of ideology and utopia. It also reaffirms its cognitive dimension of Aristotelian derivation, albeit in the context of those peculiarities that derive from new technologies, such as the autonomy to move *in* and *between* media and the plurality of interpretation, which often raises hermeneutic and therefore necessarily ethical difficulties. Transferring to an object the name that belongs to another[117] is an imaginative process that is currently absent in AI, which still struggles to understand metaphors and often produces hallucinations.

According to Wunenburger, these thinkers have given a new direction to theories of imagination and the imaginary, emphasizing certain specific lines of thought. In his view, moral imagination represents a crucial stage or capacity in understanding and realizing what is good and it appears increasingly necessary in light of the current visual turn. Man becomes a moral person to the extent that

117 The reference can only be to Aristotle's theories on metaphor, expressed in the *Poetics* and the *Rhetoric*. In his *Poetics*, Aristotle writes that metaphor involves transferring the name that belongs to one element to another: "giving the thing a name that belongs to something else; the transference being either from genus to species, or from species to genus, or from species to species, or on grounds of analogy". The questions posed by the Stagirite, although they belong to the realm of language, can be used as a basis for certain contemporary questions concerning the ethics of the image and imagination, which we could define as "metaphorical". The image is metaphorical insofar as it transcends its own meaning and evokes other images, mental and real, new visions of the world. The metaphor is based on a kind of abbreviated similarity, on the basis of an analogical relationship. In this sense, AI unconsciously creates metaphorical images. See Aristotele, *Poetics*, Engl. transl. by M. Heath, Penguin, London 1997, 21, 57b 6-9; Aristotle, *Rhetoric*, Engl. transl. by G. A. Kennedy, Oxford University Press 2007, III, 11, 1412b 34-1413a 2. See A. Laks, *Substitution et connaissance: une interprétation unitaire (ou presque) de la théorie aristotélicienne de la métaphore*, in D.J. Furley and A. Nehemas (eds.), *Aristotle's Rhetoric: Philosophical Essays*, Princeton University Press, Princeton 1994, pp. 283-305; G. Lakoff and R. Johnson, *Metaphors We Live By*, The University of Chicago Press, Chicago 1980.

he consciously seeks to regulate his actions and his life according to the norms of good and evil.[118]

In recent decades, interest in the word 'imagination' has gradually waned in favour of the newly emerging category of the 'imaginary', meaning "that which is the effect of the imagination, which does not exist, which is not real", a category permeated by social context.[119] This century has seen the emergence of the concept of the imaginary in the modern sense, representing something collective that subliminally conditions our imagination, which is now shaped by images.[120] Imagination promotes transversal knowledge thanks to a kind of mental recomposition, it helps us to make connections between different orders of reality, those "montages" of which Aby Warburg is the spokesman in *Mnemosyne* and which remind us of the

---

118 J.-J. Wunenburger, *Philosophie des images*, cit., p. 366; Id., "L'imagination comme composante du jugement éthique?", in «Revue française d'éthique appliquée», 12, 1, 2022, pp. 118-29.

119 Cf. The noun *imaginary* originates from the Latin adjective *imaginarius* meaning "of images", referring in particular to the world of the *pictor* (*pictor imaginarius* is the figure painter, the plasticiser); it also means "that exists only in the imagination, that has only appearance, fictitious, not real"; Headword *Immaginario*, in *Il Vocabolario Treccani*, Istituto della Enciclopedia Italiana, Roma 2003, p. 768; Headword *Imaginarius*, in F. Calonghi, *Dizionario latino italiano*, Rosenberg & Sellier, Torino 1989, p. 1306.

120 In this regard, Taylor offers a precise definition of social imagery: "of the ways in which they imagine their social existence, how they fit together with others, how things go on between them and their fellows, the expectations which are normally met, and the deeper normative notions and images which underly these expectations". Castoriadis, on the other hand, defines the social imaginary as a "unceasing and essentially undetermined (social-historical and psychical) creation of figures/forms/images, on the basis of which alone there can ever be a question of 'something'". Both refer to Baczko, who understands social imaginaries as 'markers' produced by each collectivity through which it perceives itself and elaborates its goals. Cf. C. Taylor, *Modern Social Imaginaries*, Durham, London 2004, p. 23; C. Castoriadis, *The Imaginary Institution of Society*, Engl. transl. by K. Blarney, Polity Press, London 1987, p. 3; B. Baczko, headword *Immaginazione sociale*, in R. Romano (ed.), *Enciclopedia Einaudi*, vol. VII, Einaudi, Torino 1979, p. 68.

mechanisms of artificial image composition through GenAI.[121] One thinks of the artistic experiments of the aforementioned Klingemann in his series *Memories of Passersby* (2018), portraits that merge male and female figures at the same time, creating disturbing and unsettling visions.[122]

In the light of these brief remarks on the concept of imagination over time, we must ask ourselves whether, in the age of technology, we are still capable of imagining according to the dictates of the past and, therefore, of making decisions. It is worth repeating that today's technical image is the result of standardisation (in bits). The hyperbolic proliferation of images, traditional and above all technical, in every context of individual life, if it stimulates our imaginative, productive and creative capacities, at the same time it limits them, saturates them, tends to standardise and unify certain aspects.[123]

According to Guy Debord, new technologies have abolished the possibility of a clear separation between reality and imagination. We live in a 'global spectacle society' in which only spectacular and impressive images attract our attention. Imagination has therefore become more than ever an active, almost abused faculty, capable not only of reproducing reality, but also of creating it *ex novo* especially with AI.[124] Imagination shows itself in the images produced by humans and machines. As Herbert Marcuse wrote long before

121 A. Warburg, *Mnemosyne. Bilderatlas*, Dölling und Galitz Verlag, München-Hamburg 1994.

122 A. Barale (ed.), *Arte e intelligenza artificiale*, cit., pp. 75 ff.

123 Some scholars have responded in the past with the theory of the imaginal. Henry Corbin's *mundus imaginalis* refers to a real and intermediate space that is not perceptible to the senses, that is understood through the *imaginative power* of each individual, and that is accessible through a visionary imagination. For Corbin, the imaginal is a mediating realm of creative imagination, rooted in Persian spirituality, where consciousness encounters the unconscious. Cf. H. Corbin, "Mundus Imaginalis, or the Imaginary and the Imaginal," in «Cahiers internationaux de symbolisme», 6, 1964, pp. 3-26, in part. pp. 3-4.

124 G. Debord, *La société du spectacle*, Gallimard, Paris 1992, Engl. transl. by K. Knabb, *The Society of the Spectacle*, Bureau of Public Secrets, Berkeley CA 2014, p. 10.

AI, "[i]magination has not remained immune to the process of reification".[125]

As Ferrarin writes, this view refers to an imagination that is "constitutive of reality because it has its own causality: it produces effects in the world, in our emotional life. [...] practical imagination is more than a double seeing or 'seeing-in', it is rather 'seeing otherwise' because it is 'seeing-in view of'".[126] It is an imagination that requires reflection on the effects it produces on society.

This constitutive and performative aspect of imagination is even more evident in digital and artificial images. Vilém Flusser coined the term "technical imagination" or "second-order imagination" in this regard, a concept more recently taken up by Lambert Wiesing.[127]

With technology, Flusser says, imagination has become an intellectual capacity for mediation. Imaginative images constitute a mediation between the world and man. But instead of representing the world, they alter it, and with AI there is a danger that it will be 'reformulated'.

"'Imagination' means, in the strict sense, the ability to reduce the world of states of affairs to scenes and, conversely, to decipher scenes as surrogate states of affairs, to elaborate and read 'topographies', including the 'topographies' of desired states of affairs, such as the future hunt (Lascaux) or the gadgets to be produced (the blueprints, or plans drawn to scale)".[128] In other words, it is about turning the world into an image. By *techno-imagination*, on the other hand, Flusser means "the ability to cipher and decipher "techno-images", i.e. images produced by means of apparatuses. [...] this capacity is completely different from traditional imagery. The 'reading' and

125 H. Marcuse, *One-Dimensional Man*, The Beacon Press, Boston 1964, p. 254.

126 A. Ferrarin, *Un mondo non di questo mondo*, cit., p. 360.

127 F. Restuccia, *Un'immaginazione artificiale?*, cit., p. 70; L. Wiesing, "Phänomenologie des Bildes nach Husserl und Sartre," in «Phänomenologische Forschungen», 30, 1996, pp. 255-81; L. Wiesing, *Artifizielle Präsenz. Studien zur Philosophie des Bildes* (2005), Suhrkamp, Frankfurt am Main 2014, Engl. transl. by N. F. Schott, *Artificial Presence*, Stanford University Press, Stanford 2010.

128 V. Flusser, *Die kodifizierte Welt*, in Id., *Medienkultur*, cit., pp. 22-23.

'writing' of techno-images [...] make completely different demands than those made on classical images (cave paintings, mosaics, stained glass, etc.). [...] then techno-imagination, the ability to produce and use techno-images, becomes absolutely essential for life, even for survival".[129]

According to Flusser, we are illiterate in matters of *techno-imagination*, we are unprepared and uneducated: "We are not capable of sending and receiving messages about our world in the 'right' way, we are not properly 'pre-programmed' for the present; we are therefore comparable not only to illiterates in a codified world dominated by texts, but also to writers who do not know grammar or spelling.[130] Our *techno-imagination* is underdeveloped and therefore does not allow us to know how to orient ourselves among the techno-images and thus to choose consciously and responsibly in this renewed universe full of digital images. This same illiteracy makes it difficult to have an ethical use of techno-imagination, guided by appropriate, morally acceptable choices.

Flusser again states that in the techno-visual dimension "the relationship between so-called 'reality' and the symbolic system (the code) is overturned".[131] Techno-images are the result of a "manipulation of the world". Techno-images use the world, not the other way round. You cut and edit images to your liking, giving meaning to the visual and to the individual who looks at it and uses it.

In reality, these tools do not replace our imaginative capacity or our imaginative activity. If used consciously, one might argue that they could help us, in some human functions.

The ethical risk, Flusser hypothesises, lies in the "threat of an apparatus that becomes increasingly autonomous from human decisions. Only a perfectly trained techno-imagination could bring

---

129 V. Flusser, *Filmerzeugung und Filmverbrauch*, in Id., *Lob der Oberflächigkeit. Für eine Phänomenologie der Medien*, Bollman Verlag, Bensheim-Düsseldorf 1993, pp. 153-66:153.

130 *Ibid.*

131 *Ibid.*; V. Flusser, *Nomadische Überlegungen*, in Id., *Medienkultur*, cit., pp. 150-159: 156.

the apparatus back under human control".[132] And, perhaps, the knowledge of the mechanisms of apparatuses (the explainability of which in state-of-the-art AI systems remains difficult at the moment) could make human beings freer to act.

The relationship between imagination and digital image technologies is then taken up by Lambert Wiesing. He draws on Vilém Flusser's technical imagination, especially the essay *Eine neue Einbildungskraft* (1990), to theorise about a "new imagination".[133] He hypothesises an externalisation of the imagination. With computer-generated imagery, images can be modified as they appear, using simple devices and easily accessible image manipulation tools, taking the image and the imagination outside the human mind, on a screen. And we can achieve results similar to our intentions. The image on the screen can be modified at will.[134] Returning to Flusser's essay, *Eine neue Einbildungskraft*, Wiesing notes how much our way of imagining has changed with the new technologies of the visual. The imagination Flusser speaks of is in some ways related to computer graphics systems when he says that it is "as if [imagination] had moved from inside (say, from the brain box) to outside (in the computer); as if one could observe one's dreams from outside".[135] What we do not decide is a kind of amputation of our imaginative freedom and autonomy. They are tools that expand our imagination, but at the same time limit it to the point of blocking it. This is a theory of imagination that is not fully applicable to GenAI systems, because in that case the individual relinquishes control over the conception of the image. However, if we understand imagination, as Flusser does elsewhere, as a "complex intentional gesture",[136] then we can apply the concept to GenAI systems. Imagination is a mental and a practical activity at the same time. It is a gesture activated by the human through the machine. And these two gestures are mutually dependent; they are body gestures that guide the human mind in

132 V. Flusser, *Filmerzeugung und Filmverbrauch*, cit., p. 153.

133 L. Wiesing, *Artificial Presence*, cit., p. 97.

134 *Ibid.*; F. Restuccia, *Un'immaginazione artificiale?*, cit., p. 74 ff.

135 V. Flusser, *Eine neue Einbildungskraft*, in Id., *Bildlichkeit*, Suhrkamp, Frankfurt am Main 1990, pp. 115-26: 123.

136 *Ibid.*

its choice of prompts.[137] We imagine something that we know we can achieve with a certain means that conditions our imagination. A productive and reproductive imagination, to use Kant's patterns. The productive imagination produces models that orient us, that give meaning and significance to certain signs, and the more flexible they are, the more they adapt to changing times. The manipulability of technical images allows such fluidity. This was previously only possible with mental images. Once these new externalised images become part of our imagery, they renew it and interpenetrate it with the old.

When we imagine an image to be produced pictorially, we anticipate the result, whereas in photography you have to give instructions to the camera and then see the graphic output. We are bringing two different imaginations into play. In contrast, the imaginative processes at work in photography and AI seem similar. Both involve a few clicks. The individual imagines what is to come, although with photography it is easier to imagine something close to what we will have, whereas with AI these expectations can also be largely disregarded due to hallucinations. However, by refining the prompts (textual or visual), we can get as close as possible to what we want. A kind of Flusserian pre-text, a linear code designed as a kind of instruction for apparatus to read images.

We do not delegate to the machine the transcendental capacity of imagination, but the capacity to produce images. For Flusser, this activity is already imagination because it is an intentional gesture that produces meaning. The operative competence of imagination is delegated to machines, and what remains in the hands of humans must adapt to new technologies. Can the design of the result to be achieved with AI be called technical imagination? The artist Gregory Chatonsky also draws attention to the concept of the externalisation of imagination, introducing the concept of *artificial imagination*[138].

137 V. Flusser, *Gesten: Versuch einer Phänomenologie*, Bollmann, Düsseldorf 1991.

138 According to Shane Denson, computational images represent "exteriorized form of imagination". Cf. S. Denson, "Artificial Imagination," in «Cinephile: The University of British Columbia's Film Journal», 18, 1, 2024, p. 12.

For Wellner there is a *digital imagination* that offers new expressive horizons, extending the human imagination. Imagination in her view follows the development of technologies in relation to what they enable.[139] Can we understand it as a simulation of human imagination? Or perhaps an externalisation of it based on images from the human imagination? Could it be a second level of imagination like Flusser's imagination? Chatonsky, in his projects such as *Deep Dream* (*The Network's Dream*, 2015, based on CNN) or *Second Earth* (2019, based on GAN),[140] introduces the idea of unconscious dream images, a dimension of which the author is unaware. Once the image is obtained from the system, something unconscious is revealed. A kind of technological unconscious according to Vaccari, optical for Benjamin, or, again, artificial imagination for Chatonsky.[141] It is an 'imaginary other', a world parallel to our own, from which we can draw. Both the world of GenAI and that of photography, in different ways, bring to light something of the experience of this world.[142]

---

139 G. Wellner, "Digital Imagination, Fantasy, AI Art", in «Foundations of Science», 27, 4, 2022, pp. 1445-1451 and the recent G. Wellner, *Can Algorithms Imagine?* in Ead. *et al.* (eds.), *The Philosophy of Imagination. Technology, Art and Ethics*, Bloomsbury, London 2024, pp. 30-41.

140 https://chatonsky.net/; A. Somaini, "On the altered states of machine vision. Trevor Paglen, Hito Steyerl, Grégory Chatonsky", in «AN-ICON. Studies in Environmental Images», 1, 2022, pp. 91-111; Id., "Dreams, Hallucinations, Visions. Generative AI and the Long History of Surrealism," in «Paradigmi. Rivista di critica filosofica», 3, 2023, pp. 541-56. On AI art and the impact of AI on contemporary visual culture and society see A. Ackerman *et al.*, *Le Monde selon l'IA,* JBE, Monteux 2025 and another project of Chatonsky about the creation of new type of image, so-called "extinction image", Cf. B. Norton, "The Extinction Image Free", in «Cultural Politics», 19, 3, 2023, pp. 333-52.

141 F. Vaccari, *Fotografia e inconscio tecnologico*, Einaudi, Torino 1979, pp. 3-5; W. Benjamin, *The Work of Art in the Age of Its Technological Reproducibility*, cit., par. XVI; M. Shawn and S. Smithand (ed.), *Photography and the Optical Unconscious*, Duke University Press, Durham and London 2017.

142 A. Somaini, "Algorithmic Images: Artificial Intelligence and Visual Culture," in «Grey Room», 93, 2023, pp. 74-115. Particularly compelling is also the distinction theorized by Sebastian Rozenberg "between artificial phantasy as indeterminate and artificial imagination as reliant on image objects". Cf. S. Rozenberg, "From Continuous to Discrete to

More recently, Ed Finn has introduced the concept of *algorithmic imagination*. It can be understood as "imaginative thinking that attempts to prefigure possible futures", but it remains algorithmically constructed, through computation.[143] The computational space of imagination becomes the space in which algorithms 'work', and we will never know "how algorithms know what they know".[144] Rather than imaginative space, we might call it a space in which algorithmic laws apply to random external stimuli. It is not possible to predict how an individual will react, even emotionally, when confronted with certain data, and what moves, even on the basis of their own 'human' imagination, they will put into action, and how subsequent data (and subsequent behaviour) will vary in response to their feedback. But can such an algorithmic process really be called 'imagination'? If the word algorithm, derived from the name of the Arabic mathematician Muhammad Ibn Musa al-Khuwarizmi, denotes a sequence of instructions for solving a problem from a given set of data, it would seem to have no affinity with imagination, although algorithms are supposed to encourage the consumption of ideas, values, imaginaries, and to emulate precisely what they lack, i.e. human imagination. Perhaps we could recover Flusser's concept of *techno-imagination*, about the ability to decipher not only images, but signs in general.[145] Algorithmic systems, like Flusser's apparatuses, calculate how to emulate imagination in our place, but without fully succeeding. On the one hand, the imaginative imagination cannot be emulated by numerical calculations; on the other hand, the use of algorithms that choose for us and guide our choices tends to flatten our own imaginative faculty, almost to atrophy it, since it does not stimulate its exercise. However, if properly trained, our imaginative capacity can counterbalance the algorithmic *techno-imagination* that individuals need to be trained in, projecting towards alternative worlds that will increasingly be

---

Continuous – Text-to-Image Models as Limit to Indeterminate Phantasy," in «Technophany», 3, 2, 2024, pp. 1-13: 4.

143 E. Finn, *What Algorithms Want*, cit., pos. 1137.

144 *Ibid.*, pos. 3878.

145 V. Flusser, *Filmerzeugung und Filmverbrauch*, cit., pp. 153-66.

part of a hybrid reality, no less real than the 'old' and unambiguous object reality.[146]

Perhaps, as Flusser observed, the drift of human imagination is due more to the absence of *techno-imagination* than to its excessive presence. In other words, the problem is not that we have delegated the production of images to artificial *techno-imagination*, but that we have not yet adequately developed the capacity to think with and through the most advanced apparatuses. Knowing how to push them to their limits might help us understand how far we can imagine with algorithms.[147]

It is necessary to experiment with and through these tools in order to push them to produce images that are not only new, but also informative and meaningful, offering new models for understanding the world, rather than merely a (distorted) mirror of it, filled with the prejudices and biases already present in mainstream object or mass media reality.[148] They can also become tools for unmasking certain social deformations.

We can either build and train ethically oriented datasets, as Paglen and Chatonsky suggest, which would require technical skills not common to all individuals, or we can try to surprise the system – play with it – to see if we get unexpected images in response. That is, to write a provocative prompt that can put the system into crisis.

Both methods can help us train our algorithmic and human *techno-imagination* in relation to each other. Otherwise, we risk the atrophy of imagination itself. In order to counteract this, it is therefore necessary, in addition to working on ourselves and our imagination, to encourage the production of images that stimulate the imaginative process. It is then up to the subjects to use critical and creative thinking simultaneously to systematise imagery, human

146 On critical ethical and political issues, see K. Crawford, *Atlas of AI: Power, Politics, and the Planetary Costs of Artificial Intelligence*, Yale University Press, New Haven Conn. 2021.

147 V. Flusser, *Für eine Philosophie der Fotografie*, European Photography, Göttingen 1983, Engl. transl. by A. Mathews, *Towards a Philosophy of Photography*, Reaktion Books, London 2000.

148 The idea of "playing against" was recently taken up by Marc Lenot, *Jouer contre les appareils*, Éditions Photosynthèses, Arles 2017.

imagination and algorithmic techno-imagination. One could think of a kind of extended imagination, again along the lines of Finn, that extends beyond the space of human cognition, or, again, of a collective imagination that connects individual-consumers to collaboratively contain machine learning processes and systems. In the wake of Chalmers and the theory of the extended mind,[149] according to which our minds are not only in our brains and bodies, but extend beyond them, even into the physical world, into the objects we use and that collect information about us, imagination takes on a broader dimension.[150]

Individuals may seek to reorient their imaginations, which can move from subjective and punctual to intersubjective and increasingly collective, social and relational. AI, if used consciously, can perhaps foster a kind of 'augmented' imagination, merging the data transformation work that individuals and computing machines do together.[151]

Such an expanded imagination can thus help the individual to remain as such, and not to become a 'dividual', as Deleuze claimed, in figuring out what the evolution of the subject might have been in the flux of an algorithmic society.[152]As we shall see in the course of this book, an ethical-philosophical perspective may provide the conceptual resources needed to orient ourselves with awareness in the face of these new imaginative and visual possibilities.

149 A. Clark and D. Chalmers, "The Extended Mind," in «Analysis», 58, 1, 1998, pp. 7-19.

150 A. Gell, *Art and Agency: An Anthropological Theory*, Clarendon Press, Oxford 1998.

151 E. Finn, *What Algorithms Want*, cit., pos. 3789 ff.

152 G. Deleuze, *Postscript on Control Societies* (1990), in Id., *Negotiations*, Engl. transl. by M. Joughin, Columbia University Press, New York 1985, pp. 177-82.

CHAPTER 2
IMAGINING AND PREDICTING WITH (AND THROUGH) ARTIFICIAL IMAGES IN THE SOCIETY OF UNCERTAINTY

2.1. *Imagining, Predicting and Suggesting in the 'Artificial' Society*

The analysis in Chapter 1 reveals a polysemic concept of imagination, which recalls conscious and unconscious brain activities. It is intertwined with the human being's aspiration not only to conceive and produce something innovative and the fruit of *phantasìa*, but also of supposing and predicting future events across time, space and cultures.

On the other hand, this latter aspect has been present in the history of philosophical thought since antiquity, with divinatory practices and prophecies. Divinatory practices aimed at the knowledge of the future in order to control the present and intervene in it. With prophecies, on the other hand, the prophet predicts the future on behalf of the divine by manifesting his will and influencing the future[1].

However, there is a significant difference between prophecy and prediction: prophecy is detached from concrete data and from

---

1 A. Rodolfi, *Introduzione*, in G. Garfagnini, A. Rodolfi (eds.), *Profezia, filosofia e prassi politica*, Edizioni ETS, Pisa 2013, pp. 7-17. Thomas Aquinas links prophecy to vision. He regards prophecy as a veiled "supernatural knowledge," which always takes the form of a reception (*acceptio*) of something; it differs from *raptus*, that is with the ecstatic condition that produces the vision of God, which is superior to prophecy. Aquinas distinguishes between various types of prophecy. Prophecy is composed of the two moments of vision and interpretation followed by proclamation, as well as sensitive and intellectual vision; the '*visio imaginaria*', on the other hand, always occurs under conditions of alienation from the senses, as happens during sleep.

reasoning. A prediction, on the other hand, is based on logical and empirical processes. Such processes can now be simulated by AI. The task of making predictions – even about some aspects of human life and the planet – is entrusted to algorithmic systems and to those based on machine learning in particular. It is not only a matter of predicting, but also of directing people towards certain choices, anticipating the elements that may arouse interest and/or needs. According to Harari's dystopian theory, the organisms are themselves algorithms and, as such, disassembled; "Rather, emotions are biochemical algorithms that are vital for the survival and reproduction of all mammals".[2] Based on the data we convey, the algorithms will tell us who we are and will be. And if emotions get the lion's share in guiding our choices, as Kahneman states, through the so-called "heuristics of emotion" and the "halo effect", the algorithmic system further complicates our architecture of choices.[3]

First of all, let us try to understand the meaning of prediction. Certainly, man's innate desire to make predictions is intertwined with the concept of vision. It is a vision that comes before (*pre-*, prefix indicating anteriority) the time of the possible occurrence, a tool to surmise what will happen, based on clues, inferences, inductions, deductions, but also more generic hypotheses or conjectures. One

2 Y. N. Harari, Homo Deus, *A Brief History of Tomorrow* (2016), Harper Collins, New York 2017, p. 53.

3 The "halo effect" is defined as the generalisation of only some of the observable aspects of an event, people, and so on, giving a judgement, benevolent or malevolent, that takes the absent information for granted by directing it in one's favour. As Kahneman writes "The tendency to like (or dislike) everything about a person – including things you have not observed – is known as the halo effect. The term has been in use in psychology for a century, but it has not come into wide use in everyday language. This is a pity, because the halo effect is a good name for a common bias that plays a large role in shaping our view of people and situations. It is one of the ways the representation of the world that System 1 generates is simpler and more coherent than the real thing". See D. Kahneman, *Thinking, Fast and Slow* (2011), Farrar, Straus & Giroud, New York 2013, p. 81; D. Ariely, *Predictably Irrational*, HarperCollins, New York 2009.

sees in the mind and/or in objective reality something that will come or some premonitory sign of what will be.

Today, machine learning systems replace the ancient prophets, relying on an algorithmic 'vision', i.e. forecasting on a probabilistic basis. The machines 'observe' and process millions of data (alphanumeric, visual, sound) and predict further information on the basis of generalizations abstracted from the training data. That is, a statistical-probabilistic method is applied to make forecasts that are refined continuously on the ground of new data being fed into the system. A type of machine learning systems are deep learning ones, characterised by the use of neural networks fed with large amounts of data (datasets) and growing capacities in performing assigned tasks. They achieve unthinkable outcomes for human beings through opaque decision-making processes.

The most advanced AI systems base their decisions on statistical correlations extracted from the training data. In contrast, human intelligence connects phenomena and decisions also under the stimulus of emotions and empathy, which AI systems are unable to emulate.

Whereas deterministic systems merely follow instructions entered during system design and the decision-making process appears clear and the outcome predictable, in the case of probabilistic methods typical of GenAI, understanding the reasons guiding their decisions is often incomprehensible. To be intelligible, the process linking inputs to outputs must be controllable and modifiable, or else be transparent. There may be 'intransparencies' of a technical nature, because not everyone has the specific skills to understand the system, or of a legal nature, in terms of what is permitted on the basis of rules that protect the scientific economy in the national and transnational sphere. Finally, there is a procedural and semantic 'intransparency' that results from the impossibility of having all the necessary elements to achieve and understand the result. A canonical example in this sense is trying to understand how a system associates an image of a pencil with a certain probability that it is in fact the representation of a pencil. It is necessary to understand the pathways between input and output and how the properties of the pencil are categorised and how the images representing it are labelled. Computational behaviour can be explained, but when the number of variables is very large – owing to the vast dimensionality and non-linear interactions, as well as the ontological complexity and opacity inherent in these systems. – it becomes difficult for humans to understand the semantic interpretation of the entire computational process. The explanatory requirements of a human being do not align with what the system can explain and convey. Therefore, it is challenging to consider such systems as highly reliable for individuals, whether they are used to predict possible completions of a sentence or drawing, as with ChatGPT or DALL-E, or to predict human behaviour. In the visual domain, new images are generated based on the recognition of a series of images paired with captions, with which the system has been trained, serving as the foundation for new image creation. The procedure is similar to the teaching method devised by Wilson L. Taylor to understand pupils' skills and knowledge. The system was based on a text where some words were deleted, and the pupil was asked to correctly complete it by inserting the missing words. The method called "cloze test" is a prediction task, which is intertwined with text comprehension. This same mechanism is characteristic of AI.[4] This is a system taken up by the Gestalt School to emphasise the human being's tendency to complete what is missing from an image, the so-called "closure or completion principle".[5] Every perceptive act, understood as Gombrich suggests as a process of formulating hypotheses and relations (relations between colours, between forms, between the individual and the community), is a culturally polarised act. Similarly, every creative act is part of a cultural tradition that guides it, even through disruptions, as Benjamin theorized with the concept of *choc*.[6]

---

4 N. Cristianini, *Machina Sapiens*, il Mulino, Bologna 2024, pp. 28-29; P. Humphreys, "The Philosophical Novelty of Computer Simulation Methods," in «Synthese», 169, 3, 2008, pp. 615-626. On cloze test, see also: W.L. Taylor, "Cloze procedure: A new tool for measuring readability," in «Journalism Quarterly», 30, 4, 1953, pp. 415-33.

5 W. Köhler, *Gestalt Psychology: An Introduction to New Concepts in Modern Psychology*, Liveright Publishing Corporation, New York 1992.

6 E. H. Gombrich, *The Sense of Order: A Study in the Psychology of Decorative Art*, Cornell University Press, Ithaka & New York 1984; Id.,

The form of a representation cannot be separated from its purpose and the demands of the society that uses that specific visual language.[7]

Today's images also change the way we think, see, remember and imagine, suggesting new criteria for analysis and new desires. From Mitchell's anthropomorphising perspective, images want something, which could also be nothing: "What pictures want in the last instance, then, is simply to be asked what they want, with the understanding that the answer may well be, nothing at all." They question us about their meaning and aim to direct our gaze.[8]

Machines aim to emulate this human ability to respond to stimuli from images, for example with the "nudging" *à la* Sunstein and Tversky and Kahneman's System 1 and 2 theories of thought. In the case of nudging, it is an ethically oriented drive based on an architecture of choices, but one that directs us on the basis of certain biases. In the case of Tversky and Kahneman's theory, the two scholars hypothesise a System 1 (more instinctive and intuitive, hence faster) and a System 2 (more rational and reflective, hence slower). System 1, in particular, relies on cognitive shortcuts that enable optimal decision-making in situations with abundant information and limited time, such as in a family context.[9]

The use of unimaginable amounts of data, the filter bubbles[10] generated by algorithmic customisation systems and cognitive

---

Art and Illusion: A Study in the Psychology of Pictorial Representation, Pantheon Books, New York 1960. See also: S. Ginger, *Gestalt Therapy: The Art of Contact*, Routledge, London 2007; W. Benjamin, *The Work of Art in the Age of Its Technological Reproducibility*, cit.

7 E. H. Gombrich, *Art and Illusion*, cit., p. 96.

8 W. J. T. Mitchell, "What Do Pictures Really Want,?" in «JSTOR», 77, 1996, pp. 71-82: 82; Id., *What Do Pictures Want? The Lives and Loves of Images*, The University of Chicago Press, Chicago 2005; Id., *The Reconfigured Eye Visual Truth in the Post-Photographic Era*, MIT Press, Cambridge Mass. 1994; D. Freeberg, *The Power of Images: Studies in the History and Theory*, The University of Chicago Press, Chicago 1989.

9 See D. Kahneman, *Thinking, Fast and Slow*, cit.; K. Frankish, "Dual-process and Dual-system Theories of Reasoning," in «Philosophy Compass», 5, 10, 2010, pp. 914-26; R.H. Thaler and C.R. Sunstein, *Nudge*, cit.

10 See "A phenomenon whereby the ideological perspectives of internet users are reinforced as a result of the selective algorithmic tailoring of

shortcuts induce a form of radically polarised and self-referential individualism. This trend fragments society into a multitude of individual choices and interests. It is no longer newspapers that shape public opinion, but information networks, which, through the aforementioned processes,[11] often isolate individuals in their own information bubbles and exclude them from views and interests other than their own.

Predictive systems not only steer our thinking in certain directions, but also influence our consumption of goods and ideas, leading us to favour certain values and choices over others. The result is a subtle submission to what is known as surveillance capitalism.[12] In such a mass media context, artificially generated images can become fake images or deepfakes, spreading misinformation and disinformation. As we will see, in the next chapter, they can also be used as tools for cancel culture.

2.2. *Big Data and "Dataism" Between Bias and Cognitive Distortion*

As is evident, big data form the foundation of artificial image generation systems. Unlike the logical methods of the human mind, which rely on connections between a limited number of elements, big data replaces these connections with accumulations of quantitatively and qualitatively unknown data, lacking any predetermined meaning. This approach differs from the past, when iconographic sources were primarily created to understand and remember history, limited in quantity due to the technical and technological constraints

---

search engine results to individual users (as reflected in recorded data such as search history, click data, and location)". Cf. headword *Filter bubble*, in Oxford Dictionary online: https://www.oxfordreference.com/display/10.1093/acref/9780191803093.001.0001/acref-9780191803093-e-482; See also: A. Bruns, *Are Filter Bubbles Real?*, Polity Press, London 2019; E. Pariser, *The Filter Bubble: What the Internet is Hiding from You*, Penguin, London 2012.

11 *Ibid.*, p. 138.

12 S. Zuboff, *The Age of Surveillance Capitalism. The Fight for a Human Future at the New Frontier of Power*, Public Affairs, New York 2019.

of the time. Today, however, there is an effort to use both historical and contemporary digital sources to predict future events, facts, and trends through new narrative hypotheses.

So, also in the visual context, algorithmic predictions are based on visual big data, through which we can: generate images; describe images; suggest, via search engines, social media, or generic sites, images that are likely to be of interest to the user doing the searching; or even ask AI systems to tell us what a particular image represents, or provide us with images based on our queries to search engines, for example. In the first case, through artificial system generators (GAN or diffusion multimodal LLM), in the second, through systems such as ChatGPT, and in the third, through the traditional machine learning procedures.[13] Finally, in the last case, systems such as Google Lens are used, where the image can be used to reveal, for example, the identity of the person depicted, what they represent (in the medical field, for example, with imaging technology) or, in the case of even more sophisticated systems, the percentage risk that a certain biotype might commit a crime, etc.

In the case of the creation and description of an image, for example, it is interesting to see which elements are avoided, which are made opaque, which are emphasised; whether biases, distortions or inaccuracies arise (of what kind and to what extent). These are issues that we will return to later in the context of visual advertising.

It is a system that offers numerous opportunities, yet it operates without understanding the meaning behind the processes it performs. It is therefore challenging to identify algorithmic systems that can emulate the human capacity and sensitivity to detect (and avoid) discriminatory aspects.

In this infocratic (or what we might now call info-iconocratic) context, data provide opportunities for knowledge and representation, yet simultaneously disempower us. Now that information is utilized to offer services, to monitor us and to predict our behaviour, we risk being subjected to a surveillance capitalism that "reduces human

13 J. Yu *et al.*, "Machine learning for big visual analysis," in «Machine Vision and Applications», 29, 2018, pp. 929-31.

beings to consumer cattle that provide data".[14] Furthermore, computer memory is accumulative, human memory is always narrative and selective[15]. Algorithms skim the information we need to store, but this comes at a cost: the freedom to select 'our narrative'.

In fact, technology allows huge amounts of data to be collected very quickly, in real time, and then processed and correlated to produce new information. After all, big data is nothing more than the result of the interaction between humans and machines.[16] Algorithmic systems can exploit user preferences for commercial, ideological, political, and public opinion targeting purposes, often to the detriment of individuals' real interests. To counteract this, individuals should be able to use a form of "self-nudging".[17]

From an ethical perspective, AI can simplistically be considered as a substitute capacity to act, autonomously deciphering the big data it relies on at need. It quickly predicts what new visual data subjects might be interested in, thus directing the behaviour and decisions of individuals. [18] The rapid response of AI is not always ethically positive, as it can undermine – even erase – an individual's ability to responsibly consider the reasons behind their choices of images, consumption, and life decisions, by questioning their own conscience.

The risk we may run, therefore, is transforming ourselves in data. As Bodei writes, "we might end up becoming heterodirected and

---

14 B.-C. Han, *Infokratie. Digitalisierung und die Krise der Demokratie*, Verlag Matthes & Seitz Berlin, Berlin 2021, Engl. transl. by D. Steuer, *Infocracy. Digitisation and the Crisis of Democracy*, Polity Press, Cambridge 2022, p. 3.

15 B.-C. Han, *Psychopolitics: Neoliberalism and New Technologies of Power*, Verso, New York 2017, p. 67; Id., *The Crisis of Narration*, Engl. transl. by D. Steuer, Polity Press, London 2024.

16 V. Mayer-Schonberger and K. Cukier, *Big Data: A Revolution That Will Transform How We Live, Work and Think*, John Murray Publishers Ltd, London 2013.

17 On self-nudging, see S. Reijula and R. Hertwig, "Self-nudging and the citizen choice architect," in «Behavioural Public Policy», 2020, pp.1-30.

18 L. Floridi, *The Ethics of Artificial Intelligence*, cit., pp. 3-30. See also H. Simon, *The Sciences of the Artificial*, MIT Press, Cambridge Mass. 1996, in part. p. 53.

therefore need to increase our vigilance against these mental Trojan horses and stick, as it were, to a kind of self-defence manual against intrusions into the sphere of our thoughts, images and passions".[19] If we limit the space of subjective choice to the boundaries set by algorithms, we are not exercising value choice. "Strong evaluation" is based on what is good for the self, weighing possible alternatives and invoking the question of responsibility for our actions and the moral principles underlying them. The ability to qualitatively evaluate available options and make reflective self-assessments are uniquely human traits.

If the algorithm follows a utilitarian approach, a kind of 'weak evaluation', the individual – following Taylor and even earlier Frankfurt – can presuppose a sense of their own actions and choices in relation to themselves, to others, and to the way each person judges their own life in terms of wholeness.

The 'question of meaning' regarding one's actions in an algorithmic context must therefore invoke the responsibility of subjects within the limits of changes emerging from new information.[20]

The domain of visual big data has given rise to processes that are not always ethical.[21] This is partly due to the (ambiguous) status of the image itself and partly to the algorithmic processes that drive images and our choices. Journalist David Brooks coined the term "dataism" in 2013, along with the concept of "dataset bias".[22]

Regarding dataism, Brooks states that "[i]f you asked me to describe the rising philosophy of the day, I'd say it is data-ism. We now have the ability to gather huge amounts of data. This ability seems to carry with it certain cultural assumptions – that everything

---

19 R. Bodei, *Dominio e sottomissione. Schiavi, animali, macchine, Intelligenza Artificiale*, il Mulino, Bologna 2019.

20 C. Taylor, *What is Human Agency?* in Id., *Human Agency and Language. Philosophical Papers I*, Cambridge University Press, Cambridge 1985, pp. 15-44; H.G. Frankfurt, "Freedom of the Will and the Concept of a Person," in «Journal of Philosophy», 68, 1, 1971, pp. 5-20.

21 It should be kept in mind as a general consideration that the most widely used data sets are in English.

22 Today we have the ability to collect such an enormous amount of data that it is considered a transparent and reliable lens for predicting the future with algorithms.

that can be measured should be measured; that data is a transparent and reliable lens that allows us to filter out emotionalism and ideology; that data will help us do remarkable things – like foretell the future".[23]

More recently, the term has been broadened in meaning to describe what Harari called a way of thinking based on the absolute value of information flows.[24]

The importance of data has led to an ongoing transition from anthropocentrism to datacentrism. Mathematical calculations aimed at prediction have weakened the human capacity to imagine. The machine's ability to process data will lead to a kind of techno-humanism. Information networks could replace and/or alter our desires and experiences.[25] The images generated by AI are the result of such data, steeped in unconsciousness. An 'other' imagination emerges, mirroring our own society and the mass media that convey it, but it is a-emotional, veiled in blue light, and polarised.

The bias mechanisms include those generated by big data, which we could call 'fluid biases'. They change as the (fast) data flow in the network changes. We can assert that automated evaluations based on big data can influence the categorization of individuals and groups, potentially leading to their exclusion or, conversely, hyper-inclusion in various opportunities, such as professional ones.

Datasets can act as generators or intensifiers (or, conversely, diminishers) of biases inherent in artificial intelligence systems. The so-called dataset biases include the inputs that determine the visual outputs of a generator, perpetuating stereotypes present in the mind of the system's designer or user through training.

---

23 D. Brooks, "The Philosophy of Data," in «New York Times», Feb. 4, 2013: https://www.nytimes.com/2013/02/05/opinion/brooks-the-philosophy-of-data.html; S. Lohr, *Data-ism: The Revolution Transforming Decision Making, Consumer Behavior, and Almost Everything Else*, HarperAudio, New York 2015; H. Glaser and P. Wong, *Governing the Future: Digitalization, Artificial Intelligence, Dataism,* CRC Press, Boca Raton 2025.

24 Y.N. Harari, Homo Deus, cit.

25 *Ibid.*, cap. 10.

This certainly applies to datasets created *ad hoc* by individuals, who will naturally catalogue images according to their own inclinations, knowledge, and cultural references. It also applies, albeit in a slightly different way, to vast open-source datasets such as ImageNet. On the one hand, a small dataset created by one or a few individuals offers a very restricted viewpoint but demands greater awareness and effort in the selection and cataloguing of images. On the other hand, in a very large dataset constructed by countless people, each with their own personality and cultural references, we obtain a broader view of the world. However, this will inevitably reflect the biases not of a small group of people, but of an entire society.

The dataset bias can generate discrimination or significant evaluative errors. For example, devices designed to identify individuals for security purposes may penalise categories of people already discriminated against based on ethnicity, disability, or gender identity.

There are many critical aspects of some software being disseminated to establish the probability that a person may commit a crime. In particular, mention should be made of the well-known COMPAS case, developed by Northpointe, whose results are based on questions about factors such as employment, substance use, residing in a stable residence, and any previous illegal behaviour. The algorithm discriminated against black people and homeless people with higher scores.[26]

But what then is meant by bias and by bias in the algorithmic context?[27] Biases reflect a particular value judgement and a simplification of diversity and complexity. They constitute a form of

---

26 F. Lagioia *et al.*, "Algorithmic fairness through group parities ? The case of COMPAS-SAPMOC," in «AI & Society», 38, 2023, pp. 459-78; J. Angwin *et al.*, "Machine Bias. There's software used across the country to predict future criminals. And it's biased against blacks", in «ProPublica», May 23 2016: https://www.propublica.org/article/machine-bias-risk-assessments-in-criminal-sentencing.

27 O.A. Osoba and W. Welser IV, *An Intelligence in Our Image The Risks of Bias and Errors in Artificial Intelligence*, RAND Corporation, Santa Monica 2017; A. Chakraborty *et al.* (eds.), *Ethics in Artificial Intelligence: Bias, Fairness and Beyond*, Springer, Singapore 2023; M. Coeckelberg, *AI Ethics*, cit., pp. 126 ff.

data distortion and cognitive distortion, brought about by prejudices, which causes incorrect predictions. They can also be something positive, guiding the learning process by narrowing the range of possible hypotheses.

The English term 'bias', now used in multiple languages, derives from the ancient Provençal *biais*, meaning 'oblique', of uncertain etymology, semantically akin to the ancient Greek *epikársios*, also meaning oblique.

Beginning in the second half of the sixteenth century, the term 'bias' took on a broader meaning in the sense of inclination, predisposition, prejudice, [from the Latin *praeiudicium*, comp. of *prae-*, "pre-" and *iudicium,*"judgement"].

Biases can now be regarded as specific types of heuristics. From the Greek verb *heurískein*, meaning "serves to find", "to discover", it refers to a method relating to the general mental processes involved in trying to solve a problem, by successive attempts and approximations until the most satisfactory solution is found. In the algorithmic domain, it is based on the analysis of the data being processed and is therefore a more dynamic process.[28]

These are mental shortcuts that allow individuals to access data held in their memory, sometimes to make judgements and sometimes to form prejudices about what they do not know and have not experienced.

In summary, if heuristics are quick shortcuts leading to quick conclusions, cognitive biases are heuristics formed without confirmation by objective reality or without critical reflection.[29]

---

28 H.E. Simon, "A Behavioral Model of Rational Choice," in «The Quarterly Journal of Economics», 69, 1, 1955, pp. 99-118; A. Tversky and D. Kahneman, "Availability: A Heuristic for Judging Frequency and Probability," in «Cognitive Psychology», 5, 1973, pp. 207-32.

29 M. Filimowicz, *Systemic Bias: Algorithms and Society,* Routledge, London 2023; P.G. Devine, "Stereotypes and prejudice: Their automatic and controlled components," in «Journal of Personality and Social Psychology», 56, 1, 1989, pp. 5-18; K.A. Appiah, *The Ethics of Identity*, Princeton University Press, Princeton 2007.

Following Kahneman and Frederick, cognitive heuristics allows us to achieve solutions with minimal cognitive effort.[30] The concept of heuristics is linked to the limitations of the human cognitive system and its relationship with rationality, highlighting the distorting effects caused by its largely intuitive operation.

As James G. March, a student of Simon's, writes:[31]

> Heuristics are rules of thumb for calculating certain kinds of numbers or solving certain kinds of problems. Although psychological heuristics for problem solving are normally folded into a discussion of limited rationality because they can be interpreted as responses to cognitive limitations, they might as easily be interpreted as versions of rule-following behaviour that follows a logic quite different from a logic of consequence.[32]

They are certainly mental shortcuts that speed up decision-making processes, even when taking contingencies into account. Visual biases can therefore be defined as special heuristics. They are systematic deviations from rational thinking, which can result from the use of heuristics. In other words, biases are disputable side effects of heuristics, leading to distortions in judgements and choices.

---

30 D. Kahneman and S. Frederick, *Heuristics and Biases: The Psychology of Intuitive Judgment*, Cambridge University Press, New York 2002 and, before, D. Kahneman *et al.*, *Judgment under Uncertainty. Heuristics and Biases*, Cambridge University Press, Cambridge 1982; See also B. Tversky, *Mind in Motion: How Action Shapes Thought*, Basic Book, New York 2019; G. Gigerenzer, *Gut Feelings: The Intelligence of the Unconscious*, Penguin Publishing Group, New York 2007.

31 In Simon's perspective, it refers to a satisfactory approach, often used almost automatically and with a reasonable level of reliability; while also enabling the decision-maker to reach a goal they had in mind, without expending excessive time and cognitive resources. Cf. R. Rumiati and P. Legrenzi, "Euristiche: da Simon a Simon," in «Sistemi intelligenti, Rivista quadrimestrale di scienze cognitive e di intelligenza artificiale», 1, 2016, pp. 169-180; H. Simon, "Invariants of Human Behavior," in «Annual Reviews Psychology», 41, 1990, pp. 1-19.

32 J.G. March, *Primer on Decision Making: How Decisions Happen*, Free Press, Detroit 1994, p. 14.

Among the many types of bias, confirmation bias (in artificial images or in image suggestions by AI systems) occurs when a person only seeks information that confirms their own views, rather than considering all options objectively. These distortions rely on, and are conditioned by, information that is familiar, from the reference culture or, in any case, pre-existing concepts.

A similar concept is the stereotype. While bias refers to a systemic error in a cognitive process that distorts judgements and decisions in a general way, a stereotype concerns content and is a generalized belief about a group of people.

Stereotypes can therefore be considered a specific form of bias, applied to the social and cultural perceptions of groups. It is a rigid and simplified set of characteristics that members of a certain social group attribute to another group, viewing it as a single and indistinct whole, without engaging in critical reasoning or acknowledging any differences, nuances, or exceptions that may exist within that group. From an etymological point of view, the term 'stereotype' refers to an imprint (from Greek *týpos*) that flattens a multitude of human beings into a rigid unicum (from Greek *stereós*), ignoring the differences between individuals and the changes taking place. It originally represented the typographical mould from which copies of a newspaper were derived. Metaphorically, it then took on the meaning of fixed and rigid psychological characteristics, attributed through categorisation, with little inclination towards transformation.

It is true that stereotypes, like heuristics, help us navigate a complex world by saving time and guiding us towards decisions that would otherwise require a significant expenditure of resources and energy. But the problem arises when they are associated with a value, a judgement, or a consideration that can lead to discriminatory actions, giving rise to what is called prejudice.[33] If stereotypes were based solely on quantitatively measurable attributes, the question of evaluation would be simpler and fairer.[34] Walter Lippmann, one

33 On prejudice see G.W. Allport, *The Nature of Prejudice*, Addison-Wesley, New York 1954.

34 S.T. Fiske and S.E. Taylor, *Social Cognition: From Brains to Culture* (1984), Random House, New York 1988, pp. 184 ff.

of the pioneering scholars of stereotypes, used the term in *Public Opinion* as early as 1922. While he defines bias in interdisciplinary social science, he describes stereotypes as [...] "distorted picture or image in a person's mind, not based on personal experience, but derived culturally".[35]

This is an inadequate and biased way of representing the world that is difficult to question even in the face of conflicting information. Cognitive, behavioural, visual and linguistic processes make stereotypes resistant to change and, therefore, relations between human beings dense with expectations. Faced with discordant information, as Allport argues, we think of an atypical exception without questioning the stereotypical assessment.[36]

---

35 W. Lippmann, *Public Opinion*, ed. M. Curtis, Transaction Publishers, New Brunswick and London 1991, pp. 79 ff. ""Stereotype" initially referred to a printing process devised in the eighteenth century. A metal plate containing source material, the stereotype enabled mass reproduction of copies. By the end of the nineteenth century, the process was beginning to be used metaphorically to describe repetitious behaviour. By the end of the twentieth century, social psychologists had redefined it as "cognitive structures that contain our knowledge, beliefs, and expectations about a social group". For the history of the concept of stereotype, see also Z. Kunda, *Social Cognition: Making Sense of People*, MIT University Press, Cambridge Mass. 1999, p. 315; S.R. Schroeder, "Usage of Stereotypy as a Descriptive Term," in «Psychological Record», 20, 1970, pp. 337-342; L.S. Newman, "Was Walter Lippmann interested in Stereotyping?: Public Opinion and Cognitive Social Psychology," in «History of Psychology», 12, 1, 2009, pp. 7-8.

36 G.W. Allport, *The Nature of Prejudice*, cit. See also: D. Katz and K. Braly, "Racial Stereotypes in One Hundred College Student," in «Journal of Abnormal and Social Psychology», 28, 1933, pp. 280-90. By highlighting how such stereotypes stem from the mass media, they are based on the practice of defining before observing, a priori. Thanks to their studies, the stereotype began to be understood as a mental shortcut that influences an individual's perception, leading them to obscure reality with biased inferences. Cf. H. Tajfel, "Cognitive Aspects of Prejudice," in «Journal of Social Issues», 25, 4, 1969, pp. 79-97; S.T. Fiske *et al.*, "A model of (often mixed) stereotype content: Competence and warmth respectively follow from perceived status and competition," in «Journal of Personality and Social Psychology», 82, 6, 2002, pp. 878-902.

Fundamentally, it is "an essential tool for processing information while economizing on scarce cognitive resources",[37] easy to remember and difficult to eradicate because stereotypes become embedded in society over time. Artificial Intelligence absorbs them from training data, propagates them rapidly and interstitially according to procedures impossible to control. Following Fiske, this scheme makes reality more easily interpretable, subject to time and historical context.[38]

In image-generative AI, a framework for the automatic detection of the most common biases in multimodal generative LLMs is needed to try to limit their social implications and the conditioning of human behaviour. An exploration of the discriminatory biases inherent in the generated images and their connection to specific visual characteristics (e.g., colour, ethnicity, gender, setting) could also be a possibility to investigate in this detection. Basically, the system could generate Alerts that anticipate the presence of possible biases to watch out for. These could be ethically oriented alerts on the non-discrimination of persons, things, cases or situations. Alerts would compel us to pay attention and engage in meta-reflection, potentially leading to a decrease in the propagation of bias and a consequent virtuous cascade in society.

Biases risk becoming dangerous prejudices without a data ethic that safeguards the dignity of human beings (and living beings in general), their physical, psychological, emotional, social, and financial security, and avoids disparities and inequalities.

To date, there are still few studies on the biases inherent in the artificial image generation process and the ethical implications associated with theme. Particularly interesting are the studies by some groups of scholars who have analysed the outcomes of certain image generation systems with regard to bias.[39]

---

37 W.P. Bottom and D.T. Kong, ""The Casual Cruelty of Our Prejudices": On Walter Lippmann's Theory of Stereotype and its "Obliteration" in Psychology and Social Science," in «Journal of the History of the Behavioral Sciences», 48, 4, 2012, pp. 363-394: 363.

38 S.T. Fiske and S.E. Taylor, *Social Cognition,* cit.; S.T. Fiske *et al.*, "A Model of (often Mixed) Stereotype Content", cit.

39 In particular, see: A.S. Luccioni *et al.*, "Stable Bias: Analyzing Societal Representations in Diffusion Models," in «Science Computers and

The fact remains that machine learning models that convert textual input into images are now widely available online and used by millions daily to generate massive volumes of images each day. Therefore, bias effects are increasingly prevalent by and for the masses. It turns out to be important to pay attention to the rapid proliferation and the growing risks caused by these models.

Stereotypes frequently emerge even from neutral prompts that merely reference descriptors, characteristics, occupations, or objects without overt social framing. Despite technical mitigation efforts (e.g., guardrails), individual attempts to elicit counter-stereotypical outputs and institutional regulations have failed to disrupt the system's structural reproduction of bias. In this light, even a best-practice prompting vademecum – while helpful in promoting ethically neutral language to avoid identity descriptors, for example, – remains insufficient to address the issue decisively[40].

Polarisations often emerge despite not mentioning specific ethnic identities or groups. The ethical impacts of today's models in society are obvious. As Bianchi writes, "A large, long-standing body of psychology literature shows when people are repeatedly exposed to stereotypical images – whether these images are real or fake – discrete social categories are reified, and these stereotypes predict discrimination, hostility, and justification of outright violence against

---

Society», preprint arXiv:2303.11408, 2023, A.S. Luccioni, "Stable Bias: Evaluating Societal Representations in Diffusion Models," in «NIPS '23: Proceedings of the 37th International Conference on Neural Information Processing Systems», 2458, pp. 56338-56351; F. D'Isa, *La rivoluzione algoritmica delle immagini. Arte e intelligenza artificiale*, Luca Sossella Editore, Milano 2024; F. Bianchi *et al.*, "Easily Accessible Text-to-Image Generation Amplifies Demographic Stereotypes at Large Scale," in «FAccT '23: Proceedings of the 2023 ACM Conference on Fairness, Accountability, and Transparency», 2023, pp. 1493-1504; A. Bondielli *et al.*, "Mining Gender Biases in Text-to-Images Language Models," in «LaGendA Workshop onGender-inclusive language in a multilingual Europe», Venice, 3-4, October 2024; V. Thapliyal and P. Thapliyal, "AI and Creativity: Exploring the Intersection of Machine Learning and Artistic Creation," in «International Journal for Research Publication and Seminar», 15,1, 2024, pp. 36-41.

40 F. Bianchi *et al.*, "Easily Accessible Text-to-Image Generation", cit.

stereotyped peoples; for example, images encoding stereotypes of Black masculinity are shown to invoke anxiety, hostile behaviour, criminalisation, and increased endorsement of violence against people perceived as Black men"[41]. Or, as Ferrara writes: "[w]hen prompted to generate images of CEOs, these models predominantly produced images of men, reflecting gender bias. This bias mirrors the underrepresentation of women in CEO positions in the real world. Furthermore, when prompted to generate images of criminals or terrorists, the models output overwhelmingly more people of colour".[42]

The difficulty in clearly mitigating them also lies in the fact that the choice of image is often dictated more by aesthetics, spectacle and the speed with which a result is achieved and propagated (downstream) rather than by values and ethical principles as the basis for the choice. Han writes, in fact, that we live subjugated by a "specific politics of visibility". From a regime of spectacle, we have moved into a regime of spectacularised and aestheticised surveillance.[43] The development of ethical guidelines, regulatory frameworks (such as the Regulation (EU) 2024/1689 on AI, so-called AI Act) and interdisciplinary research groups, "can promote more fairness, transparency, and accountability in the development and use of AI systems".[44]

However, the images generated can also raise awareness of risks in uncertain contexts. This awareness could motivate protective and,

---

41 *Ibid.*; See also D.M. Amodio and P.G. Devine, "Stereotyping and Evaluation in Implicit Race Bias: Evidence for Independent Constructs and Unique Effects on Behavior," in «Journal of Personality and Social Psychology», 91, 4, 2006, pp. 652-61; M. P. Slusher and C. A. Anderson, "When Reality Monitoring Fails: The Role of Imagination in Stereotype Maintenance," in «Journal of Personality and Social Psychology», 52, 1987, pp. 653-62.

42 E. Ferrara, "Fairness and Bias in Artificial Intelligence: A Brief Survey of Sources, Impacts, and Mitigation Strategies," in «Sci», 6 1, 3, 2023, p. 4.

43 B.-C. Han, *Infocracy. Digitisation and the Crisis of Democracy*, cit., p. 3.

44 E. Ferrara, *Fairness and Bias in Artificial Intelligence*, cit.; B. Giovanola, S. Tiribelli, "Beyond bias and discrimination: redefining the AI ethics principle of fairness in healthcare machine-learning algorithms", in «AI & Society», 38, 2023, pp. 549-563

where necessary, corrective actions.[45] This is particularly important to mitigate the frequency of images generated and selected for their aesthetic qualities—an aesthetic still influenced by the tastes of the prevailing culture. For instance, a prompt like "attractive person" tends to generate portraits closer to the Caucasian ideal rather than the African-American or Asian ideal, indirectly suggesting that individuals who do not align with this ideal are less attractive.[46]

In the case of occupations, however, as Bianchi states, in "descriptors that have comparable real-world statistics across demographic groups, such as occupations, we find cases of near-total stereotype amplification. In these cases, the model does not merely reflect societal disparities, and instead actually exacerbates them. For example, 99% of the generated software developer images are represented as white according to a pre-trained model, while in the country where the foundational training dataset was constructed (the U.S.), only 56% of software developers identified as white".[47]

Perhaps, rather than merely attempting to limit the generation of biased images, what is needed is a code of ethics for the 'good' use of the generated results and/or an alert system highlighting the main biases to watch out for, or even a bias detector, though it too could be subject to bias.

At this point, what becomes decisive is not only the quality of the generated image but also the emotional reaction of the viewer. As Raney writes, "[e]ach of these emotional reactions are regulated by morality". Moreover, aesthetic pleasure is also intertwined with moral intuition, moral emotions, and moral disengagement. [48]

---

45 D. Zillmann, *Empathy: Affective Reactivity to Others' Emotional Experiences*, in J. Bryant and P. Vorderer (eds.), *Psychology of Entertainment*, Lawrence Erlbaum Associates Publishers, Mahwah 2006, pp. 151-181.

46 A. Kardiner and L. Ovesey, *The Mark of Oppression: A Psychosocial Study of the American Negro*, W.W. Norton and Company, New York 1951.

47 F. Bianchi *et al.*, "Easily Accessible Text-to-image", cit., pp. 1494-95.

48 A.A. Raney, "The Role of Morality in Emotional Reactions to and Enjoyment of Media Entertainment," in «Journal of Media Psychology», 23, 1, 2011, pp. 18-23.

In attempting to 'interpret' an image, numerous implications and steps emerge, including logical, cognitive, semantic, aesthetic, ethical, relational and social ones. If the image is artificial, unexpected illogical elements can emerge, which we try to make sense of, even when they do not make sense.

Despite the apparent ability to generate anything imaginable or unimaginable, these models often produce images that align with dominant stereotypes, through high-profile mosaics, even when asked to produce the opposite.

The intersection of visual data, the composite nature of linguistic biases, the constant evolution of model development, the fact that they are tested and implemented in an opaque context, make quantitative and qualitative analysis of the most common biases extremely complex. For this reason, finding optimal mitigation solutions becomes quite difficult.

The accessibility and increasing use of these models – for stock photos, cheap advertising, or quick information –, combined with the silent objectification and stereotyping of certain social categories, create a dangerous mixture.

Currently, the greatest risks lie in the distorted (and often offensive) representation of women, certain groups of people, the concept of family, and various professions.

As Bianchi observes, "the output images reinforce heteronormative social institutions, which presume that marriage and traditional family structures are based on different-sex couples, contributing to the phenomenon of minority stress: those with LGBTQ+ identities disproportionately experience stress and other mental health consequences as a result of homogenizing stereotypes, stigma, and discrimination".[49]

49 Cfr. F. Bianchi *et al.*, "Easily Accessible Text-to-image," cit., p. 1496; Y. Wu *et al.*, "Image-Perfect Imperfections: Safety, Bias, and Authenticity in the Shadow of Text-To-Image Model Evolution," in «ACM Conference on Computer and Communications Security», October 14-18, 2024; S. Ghosh and A. Caliskan, *'Person' == Light-skinned, Western Man, and Sexualization of Women of Color: Stereotypes in Stable Diffusion*, in H. Bouamor *et al.*(eds.), *Findings of the Association for Computational*

In relationships of occupational disparities, even generic prompts for occupations without specifying gender or ethnicity can lead models to reinforce certain recurring stereotypes. For instance, roles such as CEO or director often skew towards male representations, while activities like childcare, nursing, and flight attendant roles tend to amplify female representations. By contrast, a moral responsibility should be recognized to refrain from reinforcing existing forms of injustice and systemic disadvantages faced by already marginalised communities. What is dangerous is not so much the creation of certain images, but their use, and the fact that the prejudices embedded in the image can more quickly than in the past shape the way we think, make choices, and perceive our world. We cannot expect users to always be attentive and aware, nor can we control the development and the implementation of these models, which so far offer endless possibilities for conveying subliminal meanings.

Cultural prejudices, moreover, depend on a number of features not only linguistic (pertaining more properly to semantics and syntax), but also on the human being's visual system and the machine's communicative system. However, we can attempt to weaken stereotypes in two ways. The first lies in their (complex if not perhaps impossible) mitigation in objective reality, the second in foregrounding the ethical impacts of generative models on individuals' choices and ways of imagining. In order to attempt to mitigate stereotypes, one must try to proceed in parallel, in offline and artificial reality; there must be a kind of repeated and widespread inconsistency in the face of certain fixed categories so as to reinforce different options and weaken the stereotype to the point of elimination. This leads one to reflect on the most disruptive and recurring stereotypical aspects and try to thwart them by showing, through GenAI, different (and non-rigid) visual alternatives as equally 'typical' possibilities. That is, it may be useful to provide alternative models to stereotypical beliefs to give value to what we may not easily remember, but which, through redundancy, may

---

Linguistics: EMNLP 2023, Association for Computational Linguistics, Singapore 2023, pp. 6971-85.

not be forgotten.[50] The ethics of communication, – understood as a reflection on the principles guiding communicative action through technologies – in this respect, can become a 'tool' to motivate a responsible and fairer use of image-generating AI; it can also lead, as a consequence, to greater trust in artificial technologies so that we can continue to develop them centered on the human being (and on living beings in general) and in a morally oriented key based on co-responsibility.

2.3. *(A)moral Strategies and AI Image Generators*

2.3.1. *Behavioral Advertising and 'Nudges'. The Human Being's Freedom to Choose, Through Artificial Images and 'Artificial' Influencers'*

Artificial images can be used in a multitude of contexts, from schoolbooks to picture books, from advertising to cinema to photojournalism, with significant ethical consequences that impact both individuals and communities. The field of investigation is certainly very broad, so we will focus here on artificial images used in two specific contexts: advertising and journalistic information.

In the universe of advertising, algorithms designed to generate new needs and incentivise the consumption of ideas, values, goods, and services motivate certain choices (and sometimes purchases) through specific machine learning models, such as the profiling of

50 See P.G. Devine, "Stereotypes and prejudice," cit.; P.G. Devine and L.B. Sharp, *Automaticity and Control in Stereotyping and Prejudice*, in T. D. Nelson (ed.), *Handbook of Prejudice Stereotyping, and Discrimination*, American Psychological Association, Psychology Press, Hove 2009, pp. 61-87; L. Marinucci *et al.*, "Exposing Implicit Biases and Stereotypes in Human and Artificial Intelligence: State of the Art and Challenges with a Focus on Gender," in «AI & Society», 38, 2023, pp. 747-61; P. Bordalo *et al.*, "Stereotypes," in «Quarterly Journal of Economics», 131, 4, 2016, pp. 1753-94.

audiences based on their online behaviour[51] and the aforementioned filtering systems.

Machine learning models automatically process incoming and outgoing data (aural, visual, verbal, etc.) provided (in)voluntarily by users on the basis of their navigation paths (and times), allowing the system to learn the characteristics of the subject and, therefore, to evolve its visual advertising proposals. The systems thus simulate the individual's cognitive capacity and make predictions about possible interests and tastes.[52]

As Thaler and Sunstein write, the new "architects of choice" are algorithms capable of reshaping and "architecting" the contexts and environments in which we form our tastes and make our choices and in which, moreover, we shape (or invent in the case of *deepfakes*) part of our personal identity.[53]

Each individual is profiled online, emerging as a digital identity different from the one that exists in the reality of face-to-face relationships. It consists of the sum of all the traces left on the web voluntarily or involuntarily during one's online actions and the traces left by other individuals. Identity is also shaped by the information inferred from the interweaving of data on similar preferences, characteristics and habits through recommendation techniques, *content-based filtering*, based on the item's characteristics, and *collaborative filtering*, based on the user's behavior in relation to others.[54]

Everyone's digital identity is part of the image we wish to give of ourselves and our relationships, but it is above all the result of a co-

---

51 Profiling "means any form of automated processing of personal data consisting of the use of personal data to evaluate certain personal aspects relating to a natural person, in particular to analyse or predict aspects concerning that natural person's performance at work, economic situation, health, personal preferences, interests, reliability, behaviour, location or movements". See EU 2016/679, Art. 4, 4: https://eur-lex.europa.eu/legal-content/EN/TXT/PDF/?uri=CELEX:32016R0679.

52 F. Pasquale, *The Black Box Society*, cit.

53 R.H. Thaler and C.R. Sunstein, *Nudge*, cit.

54 S. Tiribelli, *Moral Freedom in the Age of Artificial Intelligence*, Mimesis International, Milan-London 2022, pp. 67 ff; Ead., *Identità personale e algoritmi. Una questione di filosofia morale*, Carocci, Roma 2023.

construction that is subject to constant change. This carries the risk of constructing multiple identities and an excessive fragmentation of the self, leading to a real dissociation between one's virtual identity and that of objective reality. It actually shapes more profiles than identities in the philosophical sense.[55] Algorithmic profiling aims to define who we are on the basis of past choices and actions, published photos, tag, voice sensing, tracked online (and offline) movements, but also who we are likely to be and how we will behave, providing an estimate of our future identity and behaviour. Our identity is therefore becoming increasingly fragile and subject to change based on algorithmic suggestions.[56]

In relation to these processes, from an ethical point of view, numerous questions arise concerning the use of the artificial visual both as profiled data and as returned data to induce consumption.

With regard to profiled personal data, it is difficult to understand its scope and what the individual (in)consciously makes available. Images of oneself, of one's experience on the web, sharing events,

55 See Z. Bauman, *Identity. Conversations with Benedetto Vecchi*, Polity Press, London 2004. The philosopher wrote that "'identity' is revealed to us only as something to be invented rather than discovered; as a target of an effort, 'an objective'; something one still needs to build from scratch or to choose from alternative offers and then to struggle for and then to protect through yet more struggle – though for the struggle to be victorious, the truth of the precarious and forever incomplete status of identity needs to be, and tends to be, suppressed and laboriously covered up" (*Ibid.*, pp. 15-16). Maintaining a consistent and firm identity is challenging for two primary reasons: first, because individuals always belong to a community of shared values and ideas; second, due to the increasing fluidity and vulnerability of contemporary society. Only through relationships can we stay more connected to our authentic selves.

56 On this theme Cf. also S. Milano *et al.*, "Recommender systems and their ethical challenges", in «AI & Society», 35, 2020, pp. 957-67; T. Kant, *Making it Personal: Algorithmic Personalization, Identity, and Everyday Life*, Oxford University Press, Oxford 2020; B.D. Mittelstadt *et al.*, "The Ethics of Algorithms: Mapping the Debate," in «Big Data & Society», 3, 2, 2016, pp. 1-21; C. O'Neill, *Weapons of Math Destruction*, Penguin, London 2016; S. Tiribelli, *Identità personale e algoritmi*, cit.; J. Cheney-Lippold, *We Are Data: Algorithms and the Making of Our Digital Selves*, New York University Press, New York 2018.

places, products or other items of interest, indirectly offer insights into one's circle of contacts with similar interests. After all, every individual is, as mentioned by Floridi, an "informational entity".

Moreover, image profiling and image generation mechanisms are based on picture-caption pairs; therefore, the language chosen to define and describe a published image or one that portrays us also becomes crucial. Labelling a picture with a person's first and last name, for instance, may raise privacy issues. It becomes a form of "freedom from" those who want to intrude into our intimate sphere, individuals or algorithms. Violation of privacy jeopardises the individual's right to be the master of their personal information and its flow.[57]

In this regard, we cannot fail to mention, at least with regard to Europe, the EU Regulation 2016/679, where we read the importance of the use of one's own data and the context of its use:

> The processing of personal data should be designed to serve mankind. The right to the protection of personal data is not an absolute right; it must be considered in relation to its function in society and be balanced against other fundamental rights, in accordance with the principle of proportionality. This Regulation respects all fundamental rights and observes the freedoms and principles recognised in the Charter as enshrined in the Treaties, in particular the respect for

57 In relation to the concept of privacy: "[T]he protection afforded to thoughts, sentiments, and emotions [...] is merely an instance of the enforcement of the more general right of the individual to be let alone. It is like the right not to be assaulted or beaten, the right not to be imprisoned, the right not to be maliciously persecuted, the right not to be defamed [or, the right not to be kidnapped, my addition]. In each of these rights [...] there inheres the quality of being owned or possessed and [...] there may be some propriety in speaking of those rights as property. But obviously, they bear little resemblance to what is ordinarily comprehended under that term. The principle [...] is in reality not the principle of private propriety but that of inviolate personality [...] the right to privacy, as part of the more general right to the immunity of the person, [is] the right to one's personality". See S. Warren and L.D. Brandeis, "The Right to Privacy," in «Harvard Law Review», 193, 4, 1890, pp. 31, 33. Cited in L. Floridi, "Four challenges for a theory of informational privacy," in «Ethics and Information Technology», 8, 2006, pp.109-19.

> private and family life, home and communications, the protection of personal data, freedom of thought, conscience and religion, freedom of expression and information, freedom to conduct a business, the right to an effective remedy and to a fair trial, and cultural, religious and linguistic diversity.[58]

By personal data the EU also means "any information relating to an identified or identifiable natural person ('data subject'); an identifiable natural person is one who can be identified, directly or indirectly, in particular by reference to an identifier such as a name, an identification number, location data, an online identifier or to one or more factors specific to the physical, physiological, genetic, mental, economic, cultural or social identity of that natural person".[59]

The concept of personal data is dynamic. An isolated piece of information (in the form of a word, a static image, or an (audio)video) may not be capable of leading to the identification of an individual. However, the fact that this information can be used to identify someone by cross-referencing it with other data determines its nature as personal data. Advertisers use various tracking techniques to distinguish individuals among online users. These methods do not physically identify the person, but the browser or device used to surf the Internet. Information such as cookies and advertising IDs are also considered personal data.

Regarding algorithms that respond to individuals by providing images, it is necessary to investigate what decisions (and to what extent) we consciously and/or unconsciously delegate to artificial intelligence in the context of consumption and purchasing choices (of information, imagery, values, goods, and services). We must also consider what margins of autonomy the individual may still have, trying to preserve their critical capacity in the face of strategies based on algorithmic force aimed at directing people's preferences. If the idea of the neutrality of AI, as a mere result of computational calculations, now seems outdated, we need to explore whether and how AI can close in on clusters or 'preference bubbles' or, on

58 Cf. https://eur-lex.europa.eu/legal-content/EN/TXT/HTML/?uri=CELEX:32016R0679.

59 Cf. *Ibid.*, art. 4.

the contrary, allow for greater moral involvement. This could be achieved by incorporating ethical values proper to the social sphere of use into its system or by facilitating individuals to weigh what can be considered good or bad for themselves and/or the community based on "strong evaluations."

The profiling of our online behaviour is increasingly pushing behavioural advertising. The risk of being heterodirected through these procedures is significant. Individuals may find themselves falling into clusters that are also influenced by behavioural biases.[60]

Another mechanism unique to AI consists of the so-called "nudge" *à la* Sunstein. According to this concept, decisions are made not based on people's rationality but on the irrational component that is "nudged" according to an ethically directed choice architecture. This nudge does not prohibit alternatives, but suggests possibilities aimed at improving the subject's well-being, without forcing or restricting.[61]

To be effective, a nudge acts on the environment in which people make decisions, to make a particular choice more appealing, change the order in which options are presented, or provide information so that people are more inclined to choose more advantageous and desired options.

First, attention must be drawn to the fact that, as anticipated, behavioural advertising and nudge systems learn from data that is often permeated by our prejudices. They provide information and predictions on what might be of interest to us within a society focused on self-satisfaction, fulfilling both authentic and artificial needs, and promoting consumption.[62]

---

60 K. Dowling *et al.*, "Behavioral Biases in Marketing," in «Academy of Marketing Science», 48, 2020, pp. 449-77.

61 Nudges can be of various types. Among the most common we cannot fail to include informational nudges, which give information to help make the best decisions; default nudges, which help guide people to a more advantageous decision; social and choice architecture nudges, which change the positions of choice options to favour one direction over another. Cf. R.H. Thaler and C.R. Sunstein, *Nudge*, cit.

62 M. Bertolaso and A. Marcos, *Umanesimo tecnologico*, Carocci, Roma 2024.

Such profiling enables the creation of visual content tailored to specific audience niches, reflecting individuals' preferences, tastes and needs from ideological, cultural and commodity perspectives. What is certain is that such behavioural advertising may raise issues of overly precise profiling and, thus, privacy and even copyright issues. Profiling of networked individuals occurs through the labelling of networked users. It can be, for instance, psychometrically based, making it possible to predict certain behaviours and to develop *ad hoc* circuits of (in)controlled obsolescence, 'rationally' capturing 'irrational' needs, frailties, and tendencies that drive the subject's actions. Preferences and choices are addressed in real time, in relation to the individual's changing online searches. Thus, not only goods are exchanged, but data, values and imaginaries, which opens up questions about privacy and broadens the concept of responsibility.

Based on this process, algorithms identify relevant content for certain categories of people, based on the profile assigned to each of them. These profiles can be seen as "identity labels" that attempt to describe and classify users as belonging to specific groups, as they share common interests (cultural, religious, political, civic, social, consumer, etc.) with the assumption that these may be repeated in the future. The profile attributed to each user with the algorithms aims to summarise their identity. In this way, a probabilistic identity of the user is constructed – not as they are, but as they might become – through algorithmic interpretation of personal data, framed by the logic of prediction.[63]

Profiling, a model grounding online advertising, produces advertising that is micro-targeted to the characteristics and tastes (and changes therein) of individual people.[64] Social media uses such strategies to make certain product images appear in news feeds. These are images that can lead directly to the site of purchase, ensure the registration of users to various sites, which in turn provide new data, or attract because they respond to (presumed) needs and desires.

---

63 S. Tiribelli, *Identità personale e algoritmi*, cit., p. 63.

64 E. Pariser, *The Filter Bubble*, cit.; C.R. Sunstein and R. Cass, "Algorithms, Correcting Biases," in «Social Research: An International Quarterly», 86, 2, 2019, pp. 499-511.

As in the case of the Covid-19 pandemic, the images play on the emotionality and vulnerability of the subject, that is the uncertainty of the moment and the fragility highlighted by profiling.[65]

After all, images are today the main communicative and promotional vehicle for advertisements, whether artificially generated, digital or analogue. What ethical questions can they raise?

First and foremost, these tools allow for the reduction of costs and time for the conception and realisation of advertising campaigns. If a campaign does not achieve its planned goals in terms of visibility and sales, it can be quickly modified or transformed. Professionals working in this field are more like supervisors rather than creators.[66] It is possible to experiment with different styles depending on the audience and the type of message and content, transcending the most common expressive boundaries, at the risk of greater standardisation and homogenisation at the expense of that ideational and imaginative uniqueness proper to the human being.

It also emerges that aesthetics can be modulated quickly, to increase the involvement of the human being, while trying to avoid raising false expectations of advertised products. More effective messages can be packaged, reaching one's target audience and interests with extreme precision, simulating more truthful contexts that would otherwise be difficult (if not sometimes impossible) to reconstruct, such as for instance hospital settings, exotic scenarios, war, and so on. The artificial image can also 'educate' its audience or give information more in line with the advertiser's concept. However, if such prompt-driven verisimilitude is taken to extremes and is not stated, it results in falsehood, deceiving its audience and manipulating it. Brands could therefore be harmed, and a relationship of mistrust could be created between the potential audience and the manufacturer.

65 F. Lagioia and G. Sartor, "Profilazione e decisione algoritmica: dal mercato alla sfera pubblica," in «Federalismi- Focus: La disinformazione online e il ruolo degli esperti nell'agorà digitale», 11, 2020, pp. 85-110; S. Zuboff, *The Age of Surveillance Capitalism*, cit.

66 On the model of the useful and its founding father Jeremy Bentham, see A. Fabris, *Etica della Comunicazione*, Carocci, Roma 2014, pp.73-8.

The use of artificial images to generate advertising content then implies the propagation of stereotypes or the exclusion of certain categories of objects and/or individuals. In order to avoid inadequate, distorted or distant representations of the present cultural and social reality, it is possible to use AI to generate a wider range of diversity, for more inclusive advertisements and to reach more specific and thus wider audience segments.

A not entirely secondary aspect lies in the impact of the position of the image in the device, the colours used and its size according to the type of asset, the socio-economic and personal context of the subject and the imagery of reference. These aspects are intertwined with our preferences, fears and spending propensities.[67]

Besides these visual types, there are chatbots and virtual influencers. In the first case, an avatar 'artificially' answers questions posed by the user, simulating human conversations and allowing users to interact with them as if they were real. Chatbots can either generate automated responses that correspond to user requests (they are called declarative in this case) or manage full-fledged dialogical interactions tith the user (conversational). The latter are often referred to as virtual assistants that are far more interactive and personalised than declarative chatbots. They exploit NLP (Natural Language Processing) and machine learning to learn how to capture a user's preferences in order to provide recommendations even in the form of images of products and goods and even anticipate needs.

This type of engagement not only enhances customer satisfaction but also gives advertisers valuable data for upcoming advertising campaigns.

In the second case, an artificially created influencer (virtual influencer) interacts with the human being. These are computer-generated personas who are given a personality within a certain worldview and made accessible on media platforms to interest followers and influence them. Virtual influencer can offer shopping tips, suggests trends – via posts, stories, reels –, talks and chats. [68]

---

67 F. Lagioia and G. Sartor, *Profilazione e decisione algoritmica*, cit.;

68 D. Sisto, *Virtual influencer*, Einaudi, Torino 2024; V. Bachini and M. Tesconi, *Fake people*, Codice Edizioni, Torino 2020; J. Wiedemann (ed.),

If 'influencer' from the Cambridge Dictionary is "a person who has the ability to influence other people's decisions about the purchase of particular goods or services[69]", virtual influencers can perform the same function, but in a more flawless and planned manner through an Internet network and algorithms. As far as virtual influencers function as simulacra. They are digitally generated personas that imitate human presence without reference to an original; they enact emotions, construct narrative identities, and engage in social interaction, all within platforms manged by algorithmic logic. Their existence exemplifies a post-referential subjectivity, where the distinction between the real and the artificial loses importance.

With Instagram and TikTok accounts, like physical influencers, they show their lives and emotions. They collaborate with companies, sponsor products, 'participate' in fashion shows. Compared to human influencers, however, they are managed by expert from the companies they work with.

Virtual influencers are beneficial for brand reputation, and they are unlikely to generate scandals with inappropriate behaviour and threaten the image of the company.[70]The virtual influencer is capable of hyper-personalisation, adapting in real time to the consumers they

---

Digital Beauties, 2D & 3D Computer Generated Digital Models, Virtual Idols and Characters, Taschen, Koln-London-Madrid-New York-Paris-Tokyo 2001; E. Hund, *The Influencer Industry: The Quest for Authenticity on Social Media*, Princeton University Press, Princeton 2023.

69 Headword *Influencer* in Oxford Dictionary online: https://www.oed.com/dictionary/influencer_n.

70 Think of the well-known case of Brazilian-American model influencer Lil Miquela, who opened her first Instagram profile in 2016, or the more recent Italian case of the Birth-of-Venus influencer for the 'Open to meraviglia' campaign promoted by the Italian Ministry of Tourism in 2023. On Lil Miquela, cf. J. Drenten and G. Brooks, "Celebrity 2.0: Lil Miquela and the rise of a virtual star system," in «Feminist Media Studies», 20, 8, 2020, pp. 1319-23; B. Usher, "Rethinking Microcelebrity: Key Points in Practice, Performance and Purpose", in «Celebrity Studies», 11, 2, 2020, pp. 171-88; On *Venus* influencer: Artribune, *Venere Open to Meraviglia e i Meme Social*, 27 April 2023 (https://www.artribune.com/turismo/2023/04/venere-open-to-meraviglia-protagonista-memesocial-esilaranti/, accessed 17 March 2024).

interact with and to shifts in their sentiments, while also generating very inclusive forms of communication[71].

In addition, virtual influencers can be 'active' and reactive at all times, without limits of space, time, or personal needs.

Their artificial nature can be read from two different points of view, both of which are valid. On the one hand, it may suggest doubts about their use and the possibility of a real relationship of trust with their audience.[72] Moreover, despite the lower costs for photo shoots and events, creating and maintaining a virtual influencer requires a massive technological and ecological investment. On the other hand, the declaration of non-authenticity makes them even more credible. If, as Sisto writes, " human influencers" often give us this impression [of authenticity], then the truly non-existent, computer-generated ones take on a higher reality, authenticity and, therefore, credibility at the precise moment they confess to being such".[73] Fiction is manifested, not concealed. And it is proof of its authenticity. With virtual influencers, the aim is not to highlight their perfection, but their true nature. It is the real influencers who want to call themselves genuine when, instead, responding to a script and enhancing their appearance with *ad hoc* filters and lighting.

These are not only influencers with millions of followers, but also micro-influencers. They dominate market niches, acting as ambassadors more in line with the values of the brand they promote and the audience they address. They stimulate the production of 'authentic' content and improve, changing quickly based on short and medium-term audience feedback, enhancing the performance of campaigns.[74]

Despite the allure of the virtual world, traditional influencers are still very popular. Presumed authenticity, acquired and flaunted skills and trust conveyed by the influencer to the community are crucial. Influencers enable brands to come up with new ideas, improve products and/or services, but also to create customised

71 D. Sisto, *Virtual influencer*, cit., pp. 11-2.
72 A. Fabris (ed.), *Trust. A Philosophical Approach*, Springer, Cham 2020.
73 D. Sisto, *Virtual influencer*, cit., p. 33.
74 See M. Conti and P.P. Tricomi, "Virtual Influencers in Online Social Media", in «IEEE Communications Magazine», 60, 2022, pp. 1-13.

products designed especially for them. They are experts in their field and know their followers inside out.

Many areas are not yet fully accessible to virtual influencers, because they require more authenticity: for instance, social or awareness-raising campaigns are based on the influencers' real-life experience, where followers attach great importance to the experts/influencers' opinions and knowledge.

In some cases, an empathic relationship can even be established between virtual influencers and, at the limit, between chatbots such as Siri or Alexa and real subjects.[75]

The emotional connection between artificial reality and object reality takes on radical importance when the subject 'feels' he can trust an otherness that he recognises not so much as diversity, but as something other than himself.[76]

The virtual influencer learns algorithmically to establish an empathic relationship to gain trust and induce a loyalty process[77]. Moreover, the concept of empathy, Donise explains, originates in the ability to feel and understand what the other person is feeling or has felt in a certain time and space not so much towards another

---

75 See, for example, the application *Replika*, to offer a virtual friend, a guide, etc. On this see T. Xie and I. Pentina, "Attachment theory as a framework to understand relationships with social chatbots: a case study of Replika", in «Proceedings of the 55th Hawaii International Conference on System Sciences», 2022, pp. 2046-55.

76 On the concept of diversity and otherness, the literature is extensive. With Lévinas we read difference as otherness, as openness to the other; in difference lies relationship; Cf. E. Lévinas, *Totalité et Infini. Essai sur l'extériorité*, La Haye, Martinus & Nijhott 1961; Id., *Autrement qu'être ou au-delà de l'essence*, Martinus & Nijhott, La Haye 1974; A. Fabris, *RelAzione. Una filosofia performativa*, Morcelliana, Brescia 2016; M. Nussbaum, *Women and Human Development. The Capabilities Approach* (1999), Cambridge University Press, Cambridge 2001; J. Clapp, "The Ethics of Alterity and the Ethics of Care in Literary Studies," in «Educational Philosophy and Theory», 57, 3, 2024, pp. 262-73.

77 A. Donise, *Critica della ragione empatica. Fenomenologia dell'altruismo e della crudeltà*, il Mulino, Bologna 2020; M.L. Hoffman, *Empathy and Moral Development: Implications for Caring and Justice*, Cambridge University Press, Cambridge 2000.

human being, but within "the animation of the subhuman world" from plants to animals to objects.[78]

It represents the human ability to animate objects. And today, these 'animated objects' emulate human beings. The empathic relationship, therefore, between human and artificial being seems to take us back to the origins of the concept of empathy, with the addition of an 'animate object' capable of relating to the human being who generated it. The relationship between a real subject and a virtual influencer can also take place through one's virtual body (as, for example, an avatar). In both cases, once the individual has connected with the virtual influencer, the latter algorithmically tries to engage and empathise with him/her in order to maintain a connection that is as long-lasting as possible. To this end, it can modify its visual, vocal, or textual content according to the needs of the actual subject, essentially creating customised content for each human subject.

The authentic I/other relationship becomes more complex, but not impossible. How can an authentic relationship be established in this context between a self and another who lives in a virtual dimension and is also artificially processed? Or, as a borderline case, between an avatar and a virtual influencer?

Firstly, it is necessary to clarify what is meant by authenticity, and secondly, what kind of relationship is possible.

78 A. Donise, *Critica*, cit., pos. 183 in which Donise cites Geiger (1911). Cf. M. Geiger, *Sull'essenza e il significato dell'empatia* (1911), in A. Pinotti (ed.), *Estetica ed empatia*, Guerini, Milano 1997, pp. 61-94: 79. On the origin of the term empathy (from Greek *en*, "in" and *-páthos*, "emotion, feeling") the first occurrence seems to come from German *Einfühlung* coined by Vicher in the late 1800s (Über *das optische Formgefühl*, 1873), then translated into English, 'empathy' as a cast from the Greek *empátheia* in the early 1900s. The theory of empathy originated in the field of aesthetic theories for which see T. Lipps, *Ästhetik*, Voss Verlag Hamburg, 1903-06. Cf. also A. Pinotti, "Staying Here, Being There. Bilocation, Empathy and Self-empathy in Virtual Reality," in «Philosophical Bulletin», 37, 2022, pp. 147-62; L. Osler, *Taking Empathy Online*, Routledge, London 2021.

In an era in which authenticity invokes an individualist paradigm that we could link to what Taylor calls the "ethics of authenticity",[79] according to which the individual is motivated to choose and act in order to self-realise, the virtual influencer acts by attempting to 'algorithmically' facilitate this process of self-realisation. By authenticity, in particular, inspired by Trilling, Taylor means that "the moral ideal behind self-fulfilment is that of being true to oneself".[80] In the relationship with the other, the subject seeks this co-participation and, at the same time, the virtual influencer tries to maintain a relationship based on respect for the other and responding to their needs, trying not to act against their values in order to develop a sympathetic relationship.

The ethics of authenticity, understood in this way, takes into account the nature of the human being, recalling the eighteenth-century idea that the individual has an innate moral intuition of what is right and what is wrong, and that the virtual influencer must try to respect, in order not to fail in its role and make its own the moral sources that lie within us. Based on conception that "the inner voice is important because it tells us the right thing to do", there is a kind of authentic moral contract with ourselves[81] that the virtual influencer must understand and respect, if this contract is to function. So, it becomes important to be true to one's sense of being human, to act according to one's inclinations and according to what belongs to us. And the profiling and personalisation of content responds to this idea of being true to oneself, which is becoming more and more compelling in an era of increasing fragmentation of the self.

This approach is related to the affirmation of an expressive individualism, which spread especially after the Second World War, but was already present among the elites of the nineteenth century, which should not be confused with the desire to do what one wants without intrusions or hindrance, responding exclusively to the

79 C. Taylor, *A Secular Age*, Harvard University Press, Cambridge Mass. 2007, pp. 474 ff.

80 It is not to be confused with the concept of egoism, following the distinction proposed by Alexis de Tocqueville. C. Taylor, *The Ethics of Authenticity*, Harvard University Press, Cambridge Mass. 1991, p. 15.

81 *Ibid.*, p. 26.

need to fulfil one's personal ambitions.[82] And consumerism, which allows the individual to manage themselves and their personal space exclusively according to their own preferences through the purchase and consumption of what society proposes, is the manifestation of the culture of authenticity in its most deteriorating and trivialising aspects. Virtual influencers insert themselves trying to fulfil this need. "Authenticity", writes Taylor, "(A) involves (i) creation and construction as well as discovery, (ii) originality, and frequently (iii) opposition to the rules of society and even potentially to what we recognise as morality. But it is also true, as we saw, that it (B) requires (i) openness to horizons of significance (for otherwise the creation loses the background that can save it from insignificance) and (ii) a self-definition in dialogue. That these demands may be in tension has to be allowed".[83]

A positive authenticity, therefore, exists only if it is regarded as a moral ideal, arguable from a theoretical and practical point of view. Authentic, after all, from the Latin *authenticus* (itself from the Greek *authéntes*), means 'truthful, valid', but also 'original', which goes back to the freedom to be oneself. Moreover, as for the concept of moral ideal, Taylor specifies that he refers to it, "not in terms of what we happen to desire or need, but [as] a standard of what we ought to desire".[84] The virtual influencer hinges on this paradigm to attract and retain the subject and to accompany them in their life, trend and consumption choices.

The authentic relationship can only take place if there is a relationship, a dialogue between subject and virtual influencer. Martin Buber's relational philosophy can help us understand the meaning of an authentic relationship between subjects, characterised by mutual recognition. "All actual life is encounter" emphasises how human beings are embodied in a web of relationships.[85] Therefore, if such a relationship aims at an encounter and coming together on the part of

---

82 C. Taylor, *Varieties of Religion Today: William James Revisited*, Harvard University Press, Cambridge Mass. 2002, p. 84.

83 C. Taylor, *The Ethics of Authenticity*, cit., p. 66.

84 *Ibid.*, p. 16.

85 M. Buber, *Das dialogische Prinzip* (1923), Engl. transl. by R.G. Smith, *I and You*, T. & T. Clark, Edinburgh 1938, pos. 50; M. Friedman, *Martin*

the virtual influencer with the other, it induces them to decentralise in order to re-construct themselves from the relationships established between virtual influencer and subject. A relationship that Buber understands in multiple dimensions without, of course, mentioning a non-human reality, but which we can take up all the same in the pairs "*I-Thou*" and "*I-It*". "If *Thou* he said, the *I* of the combination *I-Thou* is said along with it; If *It* is said the *I* of the combination *I-It* is said along with it",[86] meaning in the first case the context of the relation, in the second of the experience. In the case of the virtual influencer, an attempt is made to emulate an I-Thou relationship,[87] but there is a risk of generating an I-It relationship, in which the ego (virtual influencer) understands the other as an object rather than as a Thou, with whom it can establish an authentic connection. In the I-It relationship, the It can be controlled and manipulated, and so becomes the object of the I. The I-It relationship, therefore, enters the realm of the useful. This is an instrumental relationship in which subjects are often objectified, strings of codes and profiles representing the individual user, whose interests are taken to be commodified. Thus, individuals can be treated as means to an end rather than as actual human beings.

Therefore, if a relationship is established between virtual bodies, an avatar of the subject existing in the real world and a virtual influencer, an empathic relationship is not necessarily established.

We do not actually have two distinct bodies, but we experience the body in two different ways. Starting from this duality, empathy is to be understood as closely related to the living in any dimension.[88]

AI has therefore revolutionised the world of advertising and the way it interacts and empathises with its audience. Evolving technology and strategies for customising and optimising communication campaigns, in addition to the aforementioned

---

Buber's Life and Work: the Early Years, 1878-1923, Dutton, New York 1981.

86 M. Buber, *Das dialogische Prinzip*, cit., p. 3.

87 Buber's "*I-Thou*" relationship constitutes a truly authentic relationship based on presence and reciprocity. The other is recognised for his or her essence, as a You. Cf. M. Buber, *I and You*, cit.

88 L. Osler, *Taking Empathy Online*, cit., p. 5.

problematic aspects, could offer a contribution in the direction of more inclusive advertising campaigns, usable by all.

The aspect of a more inclusive representation of society, I believe, could be the counterpoint to the exacerbation of bias. It can become one of ethical perspectives of AI in advertising. It is easier with *ad hoc* prompts to create advertisements that break stereotypes, promoting equality and involving all ethnicities, genders and ages. Advertising could thus promote greater social cohesion and acceptance. It might promote respect for diversity and the representation of many types of otherness, all worthy of being communicated.

If everyone can be represented, a sense of belonging and greater social harmony is fostered. The use of inclusive images (also mirroring an inclusive language, also present, for instance, in captions) not only incentivises and promotes the beauty of diversity but also enables brands to positively enhance their social perception and affirm their brand loyalty.[89]

It is a tool 'to counteract' prejudices and stereotypes, as well as expanding market possibilities and sensitivity to inclusive business.

In order to achieve inclusive results, however, it is necessary to have a more inclusive dataset to train the AI models. AI can enable the shaping of social perceptions on inclusion, and this could be a unique opportunity. If traditionally advertisements make use of models who respond to canons that are now standardised almost to the point of trivialisation, with AI it is possible to go further and respond to the needs of different genders, ethnicities, body types, ages and abilities, putting technology at the service of society through responsible and fair use.

Advertisements follow social norms and expectations and the generation of advertisements with AI runs the risk of not having enough data on certain areas, topics, subjects to be represented and audiences. The datasets underlying the generation of advertising

89 B. Ochoa, *AI-powered Advertising: Examining the Possibilities of AI in Creating Advertisements for All Persons*, Independent publishing, 2024, pos. 33, 45; M. Bajak and L. Spendel, "The Possibilities of Using Artificial Intelligence in Advertising: Theory and Practice," in «Scientific Papers of Silesian University of Technology – Organisation and Management Series», 197, 2024, pp. 9-23.

images still need to be implemented a lot before inclusion can really be a viable option. This is why even more responsible use is necessary today, trying to weaken the biases as much as possible by counteracting them. Furthermore, another inclusive aspect of AI in the advertising world lies in producing accessible advertisements, in which images can be described by audio, accessible without sight.

The development of this combination could make the advertising environment fairer, more transparent, democratic, and respectful of the human being and the human dignity.

What is certain is that today, as Bauman wrote well before today's algorithmic development, the constant dissatisfaction of contemporary man always in search of something and the diktat of radical consumption move those around the advertising world more than inclusive and responsible consumption. "Consumer life, the life of consumption, is not reduced to buying, and owning something. It doesn't even come down to the fact that we get rid of what we bought two days ago, which we were still proudly displaying yesterday. What distinguishes it the most, if anything, is that it is in constant motion. [...] It is illegitimate to feel satisfied".[90] The buzz caused by the artificial images variously arranged online and coming at us as solutions to our needs causes the subject to feel dissatisfied all the time and to try to banish from themselves the sense of pain caused by frustration at their lack of 'completeness'. Contemporary algophobia feeds on algorithms that promise instant pleasure. AI with its appealing images responsive to the contingency corroborates this mechanism.[91] The inadequacy of the self appears to be the instrument and end of the online marketplace of artificial images for purchases that immediately solicit new purchases. The duration of a good is no longer a value; in contrast, the apotheosis of the moment that deludes

90 Z. Bauman, *Homo Consumens*, Erickson, Milan 2007, p. 24.

91 B.-C. Han, *The Palliative Society: Pain Today*, Engl. transl. by D. Steuer, Polity Press, London 2021, p.1. Han explains how the neoliberal imperative to "be happy" capitalises on our desire for "self-improvement and self-optimisation" (*ibid.*, p. 11), pushing us to regulate our behaviour according to productivity standards and to conform to the diktat of possession, all while genuinely believing that we are free in our self-imposed control.

us into thinking we achieve happiness. Consumerism has become the foundation of modern life. Hochschild, in this regard, states that "[c]onsumerism acts to *maintain* the emotional reversal of work and family. Exposed to a continual bombardment of advertisements [...], workers are persuaded to "need" more things".[92]

Data are thus processed to respond to the social values of rapid consumption of information, images, imagery, goods and ideas. New expedients instil in individuals the need to act and suggest what to choose, while deluding them into believing they have greater control over their time and decisions. As they watch and act, people think about how to quickly escape the sense of inadequacy that pervades them.[93] The relationship between individuals and objects, and between individuals and themselves, is thus altered through the structure of advertising that aims to "consume through objects".[94] The online marketplace promotes an internal tension in the subject by generating the fear of being excluded and dissatisfied with one's identity, while simultaneously encouraging new needs (and thus possible solutions) to redefine it. Acting online thus becomes an escape from the self, a remedy for the distress of today's individual. Algorithms facilitate the process of obsolescence of our past in order to be reborn through new consumer goods that redefine our image and promise (momentary) security.[95]

But such processes are based on data and reference schemes that are distorted at the outset, rooted in the social imaginary of reference, and on deformed and simplified relations between individuals and society. It is an imaginary full of removals and self-interested fictions designed to reinforce the ideal self and, unconsciously, make one believe that it coincides with one's real self. Consumption becomes playful and, in the wake of Baudrillard, "the ludic dimension of

92 A.R. Hochschild, *The Commercialization of Intimate Life*, University of California Press, Berkeley 2003, p. 208 ff.

93 N. Aubert, *Le culte de l'urgence. La société malade du temps*, Flammarion, Paris 2003, p. 63.

94 J. Baudrillard, *The System of Objects*, Engl. transl. by J. Benedict, Verso, London, 1996, pp. 164-167, 172-178: 178.

95 N. Aubert, *Le culte de l'urgence*, cit., pp. 62-63.

consumption has gradually supplanted the tragic dimension of identity".[96]

No individual declares their authentic preferences online. They claim to prefer one type of object or idea, observe another and, based on algorithmic impulses, engages with yet another and, in the end, prefer it.[97] What happens, for example, in the case of "corrupted personalisation", a process that leads to a preference for something other than the genre that responds to the ideal self and the real self, instead responding to the algorithmic self. As Sandvig writes "[o]ver time, if people are offered things that are not aligned with their interests often enough, they can be taught what to want. That is, they may come to wrongly believe that these are their authentic interests, and it may be difficult to see the world any other way".[98]

In such a context, acheiropoietic images or images made by human hand are hinged and then fed into the algorithmic flow to follow the subject ever more closely and satisfy them in the shortest possible time. They are images designed to respond to the anxieties of the fragile contemporary individual, making them believe they are protecting them rather than locking them in a gilded cage. A change of course seems necessary, as well as a more inclusive visual communication that respects the specificities of each person – physical, social, economic, emotional, etc. – including through GenAI.

2.3.2 *Artificial Images and Visual Information. 'Truth Effect' Beyond Accountability, Right to Information and* Artificium

In recent years image-GenAI has also entered the world of news reporting. Although it opens up new opportunities in the creation of visual content and narrative modes of facts and events, it raises

96 J. Baudrillard, *La société de consommation*, Editions Denoël, Paris 1970, Engl. transl. *The Consumer Society. Myths and Structures* (1974), Sage Publication, London 1998, p. 192.

97 E. Finn, *What Algorithms Want*, cit.

98 C. Sandvig, "Corrupt Personalization," in «Social Media Collective», 26/6/2014: https://socialmediacollective.org/2014/06/26/corrupt-personalization/.

ethical questions regarding the concept of truth, public opinion, misinformation, and disinformation.

Image-generative AI proves useful for depicting abstract concepts, situations, events, or complex scenarios otherwise difficult to illustrate. Consider unique weather conditions, war scenarios, dangerous situations of various kinds, etc. It can also enhance an existing image, make it more effective from a communicative point of view, by adjusting its light, colour tones, etc., far beyond the capabilities of traditional computer graphics tools.

In terms of communication, image-generative AI allows for enhancing the efficiency and effectiveness of the message. It customizes the image by adapting it in the most coherent way to the journalistic article and the layout of the newspaper (print or online). It also explores new possibilities of visual representations, pushing the boundaries of journalistic storytelling. The uniqueness of the iconography of that image, like a photographic snapshot, becomes the hallmark of its originality.

Two possible questions therefore arise for our consideration: the first relates to being put in a position to understand when an image is generated by AI and to what extent. This is the criterion of transparency, which aligns with the prevailing ethic of openness today, to which it explicitly refers.[99] The noun 'transparency' derives from the medieval Latin *transparens*, composed of *trans*, 'through', and *pareo*, a verb that can mean 'to see', in a transitive sense, but also, to appear, in an intransitive sense. The same meanings are also recalled by the Greek adjective *diáphanos*, consisting of the preposition δια, 'through', and the adjective *phainómenos*, 'visible to the senses'. From the verb *phaíno*, it recalls 'to show' in the sense of 'to make seen' and, therefore, 'to make known'; in its passive form, however, the verb phainomai means 'the appearing' and 'the visible being'. That is, what is transparent acts as a filter, making it visible beyond that transparency and, at the same time, implies a mode of sight itself, which uses a filter to see. Transparency, more generally, means clarity and the absence of any desire for concealment and secrecy. To be credible one must be transparent, like a "glass house",

99 L. Floridi, *The Ethics of Artificial Intelligence*, cit., pp. 57-66.

in which everything is visible to everyone, to borrow a well-known expression of Turati.[100] Contrary to *opaqueness*, the *openness* ethics is based on unrestricted accessibility of data and transparent data, i.e. data whose origin can be recognised. This is the data from which images are then generated.

On the one hand, therefore, transparency is demanded in declaring the source of the artificial image, on the other hand, it is considered that through the network as much data as possible should be given access with a view to radical pluralism and accessibility.[101] This requires responsibility on the part of all players involved, from image producers, users and those who algorithmically process the data available.

On the principle of transparency, the European AI Act intervened forcefully in Chapter IV, *Transparency Obligations for Providers and Deployers of Certain AI Systems* (Art. 50, para. 3), which requires that "users of an AI system that generates or manipulates image, audio or video content that appreciably resembles existing persons, objects, places or other entities or events and would falsely appear to a person to be authentic or truthful ('deep fake'), shall disclose that the content has been artificially generated or manipulated".[102] An exception is foreseen when the use of such generated content is necessary for artistic expression and the exercise of free thought.

Secondly, one must ask oneself, if the representation is artificial, whether to believe the message conveyed by the image (i.e., its 'truth'). In this sense, the concept of truth needs to be re-semantised and the role of trust needs to be recovered. To quote McLuhan, "the 'content' of any medium is always another medium. The content of writing is speech, just as the written word is the content of print, and print is the content of the telegraph".[103]

---

100 F. Turati, *Atti del Parlamento italiano, Camera dei Deputati, sessioni 1904-1908, 17 giugno 1908*, n. 22962.

101 D. Resnik, "Openness in Scientific Research: A Historical and Philosophical Perspective," in «Open Access Law», 21, 11, 1, 2023, p. 132.

102 https://artificialintelligenceact.eu/.

103 M. McLuhan, *Understanding Media*, McGraw-Hill, New York 1964, pp. 23-4.

So, we have several options: we can legitimately no longer believe in anything that is offered to our gaze or – if "the medium is the message" – we can try to trace what is 'true' in the 'medium-world' within the 'artificial visual medium' according to the logic of "remediation".[104]

After all, we live in an increasingly visual information regime. An info-iconosphere that, as Han states in relation to the infosphere, has generated a "form of domination in which information and its processing by algorithms and AI have a decisive influence on social, economic and political processes".[105] Therefore, ambiguous, hyperreal, unreliable or unacknowledged visual information can give rise to arbitrary choices and moves far beyond the world of news. Information can be defined as an attempt to describe facts, situations, persons, or events as objectively as possible. This knowledge obtained from investigation, study, or instruction.[106]

Visual journalism is enhancing and expanding its role. With the widespread diffusion of the Internet, one has the possibility to visualise, in real time and wherever one is, what would previously have remained unknown.[107] The ease of multiplying and transmitting

---

104 J.D. Bolter and R. Grusin, *Remediation: Understanding New Media*, MIT Press, Cambridge Mass. 2000, pp. 20 ff.

105 B.-C. Han, *Infocracy*, cit., pos. 10.

106 Headword *Information* in Cambridge Dictionary online: https://dictionary.cambridge.org/dictionary/english/information; and in Merriam Webster Dictionary: https://www.merriam-webster.com/dictionary/information. The difference between information (understood as the unidirectional passage of a message from an issuer to an addressee) and communication (as the bidirectional passage of a message from an issuer to an addressee and vice versa), certainly appears to be more blurred today than in analogue time, almost to the point of its cancellation. In this context, however, I mean information as the domain of description of facts and events within the journalistic profession and those who deal with news. See A. Fabris, *Etica del giornalismo*, Edizioni ETS, Pisa 2019; Id., *Etica della comunicazione*, cit.; N. de la Peña *et al.*, "Immersive Journalism: Immersive Virtual Reality for the First-Person Experience of News," in «Presence: Teleoperators and Virtual Environments», 19, 4, 2010, pp. 291-301.

107 "Visual" journalism, "based on the combination of text and images and [which...] focuses on the expressiveness of the latter," was born and

images, of reshaping them to make them more effective and communicative, as well as their immateriality and immediacy are constitutive figures of a new way of communicating, more direct, immediate, fitting, intuitive and empathetic. Similarly, the speed at which an image is disseminated has changed the concept of time, which has now become a constant present. Information represents the "cosmology of contingency", and the information image aims precisely at stopping that contingency.[108]

Moreover, we live immersed in what Han calls a new praxis of the human, 'infomania'[109], developed in a multiplicity of signs: linguistic, visual, aural, gestural, etc.

A single image-information can (re)construct an event, tell a story, offering itself for interpretation. The very things of the world can become sources of information, transforming themselves into "infomes", "information-processing agents".[110] 'Things' communicate with us without even realising it. Objects that listen, that act, that respond without explicit requests or that track and advise us. Debord foresaw the algorithmic visual domain where "the real world is transformed into mere images, [and] mere images become real beings."[111] These evolving entities interact with humans,

spread in the first half of the 19th century parallel with the birth of photography. Influenced by the American *penny press* and *yellow journalism*, the illustrated press increasingly adopted the image as a means of narrative legitimation. Early instances of war photography, from Lecchi (1949) to Szathmari (1853-1856, during the Crimean War), mark the emergence of a visual paradigm in which "seeing" becomes an epistemic condition for "believing," anticipating the reflections raised by Susan Sontag about the moral and epistemological status of images. Cf. S. Sontang, *On Photography* (1977), Penguin, London 2008; V. Neri, *Notizie visive*, Pacini Edizioni, Pisa 2021, p. 26; Ead., "The Autonomy and Responsibilities of the Image in Cyberspace", in «Phainomena», 100/101, 2017, pp. 223-239;

108 N. Luhmann, *Entscheidungen in der "Informationsgesellschaft"*: https://www.fen.ch/texte/gast_luhmann_informationsgesellschaft.htm.

109 B.-C. Han, *Undinge. Umbrüche der Lebenswelt*, Ullstein Verlag, Berlin 2021, Engl. transl. by D. Steuer, *Non-Things, Upheaval in the Lifeworld*, Polity Press, London 2002, pos. 15.

110 *Ibid.*, p. 8.

111 G. Debord, *The Society of the Spectacle*, cit., p. 11.

exemplifying the informative power of images, but also their deep charm and influence on public opinion. Visual information is reshaping the traditional dominance of words in news, due to our growing inability to focus.[112]

However, further reflection is needed on this alleged visual domain. Artificial images come from textual commands and are processed and 'adjusted' on the basis of words. Therefore, if they appear predominant in the information scene, they remain the product of a skilful use of language.

Are we facing a new, subterranean linguistic turn? In my opinion, a new interaction is taking place between the linguistic sphere and the visual sphere. The artificial image can be considered as an oil painting in which each brushstroke is like a word and only the skilful sum of each brushstroke-word, with seamless corrections and nuances, can give the desired communicative result. No one sees the artist's work in his studio, his paints, preparatory sketches and turpentine to erase mistakes, but only the end result to be exhibited. The practice of capturing images instantaneously and subsequently publishing them has become obsolete. Instead, a new approach has emerged, whereby verbal descriptions are meticulously crafted to serve as a visual representation of the narrative being conveyed. Just as in a newspaper article one recounts facts, events and/or expresses a stated point of view, so does the artificial image. New skills need to be developed in order to responsibly create an effective and truthful representation.

In this sense, images again become ancillary to text even though we are immersed in the iconosphere. And perhaps, as in the past, in order to have the most complete information possible, text and artificial images can only be complementary once again.

---

112 Bruce Morton states that our brains are trying to adapt to the new contemporary mass media context. It is a context in which information comes from an increasing plurality of sources; what it requires is to train and to speed up to all these digital suggestions by metabolising them. Cf. J.B. Morton *et al.*, "Age-related Changes in Brain Activation Associated with Dimensional Shifts of Attention: an fMRI Study," in «Neuroimage», 46, 1, 2009, pp. 249-56.

The fact remains that images swiftly cater to the audience's desire for immediate (com)participation in an event and identification. After all, as Flusser maintains, the image arises from man's ability to distill meaning from fundamental elements of worldly reality, transforming lived experience and information into symbolic, visible forms that can be shared and remembered. It is an ethical act of abstraction, aimed at preserving and communicating essential aspects of reality.[113] However, within the info-iconosphere inhabited by "infomes," we have transcended digital photography by employing AI.

With the AI-generated images, we can assert that we are faced with a mechanical interpretation of the digital iconographic world, an information modality partly distorted by the reference dataset. These datasets, predominantly reflecting Western imagery and mainstream themes, have been collecting data since the 1990s, witnessing an exponential surge in the last decade with the Internet. The visual hyper-reality of some of these generated images poses some difficulties in interpretation, making it difficult to discern the true from the false without explicit declaration by the source.[114] Therefore, we can adopt a consciously critical vision or a disenchanted and nihilistic vision, according to which we no longer believe anything that we see before our eyes. Lack of trust can take over the search for what is true in the image. Images, now, hold such a level of uncertainty that the observer's search for trustworthy points can't fully counterweigh the ubiquitous lack of trust.

In the field of information, the criterion of visual truth is crucial, so much so that the European AI Act not only demands, as anticipated, that so-called deepfake images be distinguished from real ones, but also calls for safeguards to avoid the generation of false or illegal content. Just as we see the emergence of immersive journalism, which highlights the growing need for the individual to be emotionally involved in the news: knowing through experiencing.

113 V. Flusser, *Into the Universe*, cit.
114 Or if effective detectors cannot be used.

A deepfake is a process that applies artificial intelligence algorithms to produce photos, videos, and audio,[115] based on real content (images, videos and audio). After all, the term deepfake, coined in 2017, is the union of *deep*, which recalls deep learning (DL) systems, and *fake*, false, i.e. a system that employs DL algorithms to modify images, sounds and videos and generates a realistic-looking synthetic/fake media.[116]

It is a technology based on neural networks trained on large amounts of data. It is used to produce, manipulate people's faces, create and manoeuvre facial movements and expressions, facial attributes such as skin colour, age, gender, expressions, but also to alter or generate places, animals, objects, synthesise an individual's speech, etc. with extreme realism[117]. Benevolent use can be made of it, creating, for instance, explanatory tutorials, digital influencers, training videos, for health and pharmaceutical care, virtual fashion shows, e-commerce, consultancy, media and entertainment applications, reconstructing facts and events in the field of justice or to reconstruct, as with VR, historical environments and characters; but malicious approaches can take over, such as developing algorithms capable of transferring celebrity faces into pornographic videos, into acts of cyberbullying or violence more generally, to defame or propagate certain ideas and thoughts, etc.[118]

---

115 Deepfake-generating systems are typically multimodal, operating across multiple channels of perception – visual, auditory, and textual – thus raising complex ethical questions about the nature of representation and authenticity. On multimodality, cf. A. Lyons, *Multimodality*, in Z. Hua (ed.), *Research Methods in Intercultural Communication: A Practical Guide*, Wiley-Blackwell, Hoboken 2016, pp. 268-80; K.L. O'Halloran and B.A. Smith (eds.), *Multimodal Studies: Exploring Issues and Domains*, Routledge, London New York.

116 E. Meskys *et al.*, "Regulating Deepfakes: Legal and Ethical Considerations", in «Journal of Intellectual Property Law & Practice», 15, 2020, 24 ff; N. Schick, *Deepfakes. The Coming Infocalypse*, Twelve, New York 2020.

117 A. Tversky and D. Kahneman, *Judgment Under Uncertainty*, cit., pp. 585-600.

118 That is, going against the recognised ethical paradigms of beneficence and non-maleficence. Cf. L. Floridi, *The Ethics of Artificial Intelligence*,

The image can be altered to the point of deception by influencing – unconsciously – the choices and actions of human beings. Moreover, the datafication of individuals through deep fake attacks conducted without consent can breach privacy and compromise security.

These processes make the fake virtually indistinguishable from the original. The result depends on the reference big data and on the control systems.[119] It is now urgent to generate new deepfake detection algorithms to unmask instances of malicious AI use, the so called *deepfake detectors*.[120]

Such images can also have a strong emotional impact, generating alarming consequences on the viewer. Deepfakes and impactful images can become propaganda strategies, to orientate thinking in a context of uncertainty by playing on the vulnerability of the subject.

Immersive journalism, on the other hand, represents a new frontier in the realm of journalistic information.

It enables individuals to experience firsthand events or situations narrated in news programmes and documentaries through virtual reality or augmented reality technologies. This sensory immersion allows individuals to immerse themselves in the described scenario.[121]

In all these cases, we are dealing with 'intransparent' images, subject to a potentially endless transformative process – rich, moreover, in *bias* that makes it difficult to understand the meaning and, if any, the truth of the image.

The adherence to the truth of reality therefore remains the decisive question. Arendt observes that "[c]onceptually, we may call truth what we cannot change; metaphorically, it is the ground on which we stand and the sky that stretches above us".[122] So if truth therefore

---

cit., pp. 57-66.

119 See S.J. Russel and P. Norvig, *Artificial Intelligence. A Modern Approach*, Pertinence Hall, Upper Saddle River (NJ) 2021, p. 39.

120 L. Gaur (eds.), *Deepfake. Creation, Detection and Impact*, CRC Press, Boca Raton 2022, p. 91 ff.; M. Filimowicz, *Deep Fakes: Algorithms and Society*, Routledge, London 2022; N. Schick, *Deepfakes: The Coming Infocalypse*, cit.

121 N. de la Peña *et al.*, "Immersive Journalism", cit., pp. 291-301.

122 H. Arendt, *Truth and Politics*, in Ead., *Between Past and Future* (1968), Penguin Publishing, London 2006, p. 264.

seems to be something earthly, with digital and AI the era of truth seems to be coming to an end. After all, the concept of truth has been explored as fundamental throughout the history of philosophy. Plato first defined it as the 'correspondence' between discourse and things, between thought and reality. A discourse is true if it describes things as they are; it is false if it describes things as they are not in the world of objects.[123]

In AI, visual information does not stand for events, it simulates events.[124] One loses the sense of reality and, therefore, of truth. We no longer know how to relate to information, Flusser presciently argues, because we no longer have the tools to relate to it.[125] He suggests that human beings no longer function as *homo faber* but turn into *homo ludens*, devoted to entertainment, replacing hands with fingers. This shift creates a new paradigm in which reality is almost entirely influenced by entertainment and infotainment[126]. Nevertheless, there is a space for moral entertainment, which simultaneously fulfils cultural and social functions. By aligning itself with the audience's knowledge and imagination, the communication of moral order becomes more effective. Finally, Luhmann claims that "[w]hat the romantics longed for in vain, a 'new mythology', is brought about by the entertainment forms of the mass media.

123 On truth and AI cf. A. Fabris, *La filosofia nell'epoca delll'intellligenza artificiale*, cit., p. 111 ff. The author asserts the new role assumed by the addressee of truth, who becomes its witness and is called to bring truth into being.

124 In relation to the concept of truth, Fabris offers a compelling insight: "It is the one who is summoned to respond to and for what is disclosed to them. [...] It is the one who assumes responsibility for this truth they have verified, committing themselves to affirming it and bringing it into actuality. In this sense, the witness is called to bring truth into being-to make it truly be." Cf. A. Fabris, *La filosofia nell'epoca dell'intelligenza artificiale*, cit., p. 112.

125 V. Flusser, *Auf dem Weg zum Unding,* in Id., *Medienkultur*, cit., pp. 185-190.

126 B.-C. Han, *Good Entertainment: A Deconstruction of the Western Passion Narrative*, Engl. transl. by A.N. West, MIT Press, Cambridge Mass. 2019.

Entertainment reimpregnates what one already is; and, as always, here too feats of memory are tied to opportunities for learning."[127]

The crisis surrounding truth, the rampant infodemic, the personalised nature of visual data, the confinement within knowledge bubbles, and the generation and reception of images by machine learning and AI systems fundamentally challenge the authenticity and, subsequently, the reliability of visual information.

Artificial information is not yet subject to *ad hoc* regulation. In some respects, at least in Europe, it is being regulated for safe, sustainable and responsible use in order to protect the public as effectively as possible from potential harm and discrimination.

More specifically, with regard to the aspects of visual information – which are considered to be of medium to low risk[128] –, the problem of copyright for generative AI training[129] arises; restrictions on recommendation algorithms are called for, and issues related to the respect for privacy (caused by the indiscriminate collection of

---

127 N. Luhmann, *Die Realität der Massenmedien*, Westdeutscher Verlag, Opladen 1996, Engl. transl. by K. Cross, *The Reality of the Mass Media*, Stanford University Press, Standford 2000, p. 58.

128 Four levels of risk are identified in the AI Act, from highest to least impactful: 1) systems with unacceptable risk, 2) systems with high risk, 3) systems with transparency risk, and 4) systems with low or minimal risk. Relative to the risks caused by artificial images and deepfakes, it is the third level that is deputed to attempt a form of protection. Because of the difficulty in understanding whether or not they are products made by human hands, regulation imposes a requirement for transparency about their origin with a view to more reliable and, therefore, more ethical AI. See C. Novelli *et al.,* "Taking AI Risks Seriously: a New Assessment Model for the AI Act," in «AI & Society», 39, 2024, pp. 2493-97; E. Leinarte, "The Classification of High-Risk AI Systems Under the EU Artificial Intelligence Act," in «Journal of AI Law and Regulation», 1, 3, 2024, pp. 262-80; V. Neri, "Generative Artificial Intelligence, Deepfakes and Vulnerable Identity. The Ethics of Uncertainty as a Response to (In) controllable Risk," in «Teoria», 2, Edizioni ETS, Pisa, 2024, pp. 79-94.

129 The AI Act imposes an obligation to document copyrighted content used to feed the model, as well as a requirement to indicate whether the content was generated by an artificial intelligence. In contrast, with regard to protection under copyright law of works created by AI systems, the AI Act is silent, leaving the relevant regulation to national regulations as well as the work of case law.

biometric data from social media or films, from public spaces to create facial recognition databases, etc.) and the authorship of the visual emerge, which are more in the legal sphere.

Another ethically important aspect recalls the censorship of some image-generating AI systems for ambiguous requests based, for example, on prompts centred on images of violence, nudity, sexuality, etc. Finally, issues of image transparency may come into play. In the European AI Act, Article 134 reads:

> Further to the technical solutions employed by the providers of the AI system, deployers who use an AI system to generate or manipulate image, audio or video content that appreciably resembles existing persons, objects, places, entities or events and would falsely appear to a person to be authentic or truthful (deep fakes), should also clearly and distinguishably disclose that the content has been artificially created or manipulated by labelling the AI output accordingly and disclosing its artificial origin. Compliance with this transparency obligation should not be interpreted as indicating that the use of the AI system or its output impedes the right to freedom of expression and the right to freedom of the arts and sciences guaranteed in the Charter, in particular where the content is part of an evidently creative, satirical, artistic, fictional or analogous work or programme, subject to appropriate safeguards for the rights and freedoms of third parties. In those cases, the transparency obligation for deep fakes set out in this Regulation is limited to disclosure of the existence of such generated or manipulated content in an appropriate manner that does not hamper the display or enjoyment of the work, including its normal exploitation and use, while maintaining the utility and quality of the work.[130]

Image GenAI therefore return complex images, that are fraught with risk and imbued with social biases. In a context reminiscent

130 In addition, it is also appropriate to envisage a similar disclosure obligation in relation to AI-generated or manipulated text to the extent it is published with the purpose of informing the public on matters of public interest unless the AI-generated content has undergone a process of human review or editorial control and a natural or legal person holds editorial responsibility for the publication of the content.". Cf. https://eur-lex.europa.eu/legal-content/EN/TXT/?uri=CELEX:32024R1689, (134); see also: T. Hagendorff, "The Ethics of AI Ethics: An Evaluation of Guidelines," in «Minds and Machines», 30, 2020, pp. 99-120.

of Bourdieu's *habitus*, where social models of classification and perception intersect with generative models, a socially biased AI emerges, [131] influenced by prejudices regarding ethnicity, gender, profession, and age. The dominant datasets reflect the populations most active on the web. The biases are not only present in the datasets, but also in the norms, practices, and processes that underpin how AI is commissioned, developed, implemented and deployed, in the culture and society at large. The creators and designers themselves, as already mentioned, can allow their biases to permeate the whole process, from the choice of data to be used to the models to be developed.[132]

Flusser encourages us to "create something improbable from the possible. [...] Men manipulate the keys to entice automatic systems into producing unlikely scenarios in their programmes. They use the keys to allow something far-fetched to emerge from the calculated punctual universe generated by the devices. This improbable world, with its unifying ability, must envelop the punctual universe like a skin to give it meaning. The power to standardise aims to provide concrete meaning to this universe, starting from the abstract and absurd universe we find ourselves in".[133]

AI constructs a uni-formed reality according to certain standards and, at the same time, a hyper-reality that is used for information purposes. It is therefore a high-risk AI – in contrast to the risk levels listed in the AI Act – if one considers the consequences in terms of beliefs and decisions that certain 'false' images can produce.

---

131 P. Bourdieu, *Outline of a Theory of Practice*, cit., p. 85.

132 R. Schwartz *et al.*, *Towards a Standard for Identifying and Managing Bias in Artificial Intelligence*, Special Publication (NIST SP), National Institute of Standards and Technology, Gaithersburg, MD 2022; A. Luccioni *et al.*, "Stable Bias" cit. To avoid images imbued with certain biases, so-called "reverse bias" can be generated: that is, the AI makes images with a predominance of features that are generally not popular on the Web and, for that very reason, not very credible. If a prompt such as 'comely girl' returns images predominantly of women of Caucasian ethnicity, training the system to reverse the bias might generate only images of female figures of African American ethnicity.

133 V. Flusser, *Heinweg von Papier*, in V. Flusser, *Medienkultur*, cit., pp. 61-66.

False images, or images that convey false content (fake images), are difficult to detect, but most importantly, they create misinformation and disinformation with the consequent impact on society.[134] Misinformation is an umbrella term that covers different types of false information. More specifically, it refers to real information that is shared with the intent to cause harm. For example, people who are unaware that a piece of information is false may share it on social media, thinking they are being helpful. Disinformation, on the other hand, is false information that is deliberately created and/or disseminated in an attempt to mislead the public. It can cause clear direct and indirect harm, for example, by undermining public confidence. It may include information that is wholly or partially fabricated, deliberately taken out of context, exaggerated or lacking in key details. The line between misinformation and disinformation may be blurred.

For, as Harari writes, information only leads to truth in a simplistic view. Today's information networks are fundamental to understanding reality and having power over it. But they also serve to create and maintain social order, and – if manipulated and selected according to certain criteria (and thus also through misinformation and disinformation) – to consolidate power. It is clear that in such a context, absolute knowledge becomes difficult to achieve, especially when truth itself is questioned and difficult to pin down. In essence, information can be used to create a narrative that justifies certain behaviours and reinforces the existing order, even if this narrative is not necessarily true.[135]

Many high-profile journalists are now using AI to visually narrate events and facts,[136] but it remains essential to keep humans at the heart

---

134 K. Sue *et al.* (eds.), *Disinformation, Misinformation, and Fake News in Social Media: Emerging Research Challenges and Opportunities*, Springer, Cham 2020; L. Floridi, *The Philosophy of Information*, Oxford University Press, New York 2011; A. Fabris, *La filosofia nell'epoca dell'intelligenza artificiale*, cit.

135 Y.N. Harari, *Nexus. A Brief History of the Information Network from the Stone Age to AI*, Vintage Digital Publishing 2024, chap. 2, pp. 53 ff.

136 Michael C. Brown, an American photojournalist, has worked extensively in this direction: https://it.adcreative.ai/post/the-future-of-advertising.

of the information process to ensure reliability and transparency and to take responsibility for what is communicated.

We are navigating in a new, unknown information landscape. Photography, in its many declinations and transformation, will remain a fundamental tool for representing the truth (although with the introduction of digital every image is potentially fake), but its power and role in relation to emerging algorithmic technologies must be monitored. The latest systems integrated into smartphones, the first roots of which date back to 2016, can now credibly add characters, details, new scenarios to the photographs just taken.[137] Whereas a few years ago the creation of altered images required specific expertise in graphic programs, it now takes only a few seconds and minimal skills. Trust in online images seems even more of a challenge. We are all responsible for the truths we tell and read online, acting in a context where the unlimited community of communication can interact and participate in the construction of truth. Perhaps the new photojournalists should also become skilled prompters? What about ordinary people who are skilled detectors and controllers of what they see?

In conclusion, the effectiveness of artificial visual information does not depend so much on the realism of the images or media used, but rather on the symbolic, narrative power and emotional impact it manages to evoke.

In other words, to conclude the reflections on both advertising and the artificial impact of information, the algorithmic architecture of choice ends up reinforcing the phenomenon that Nicholas Negroponte already called "Daily Me" in 1995, the formation of personalised choice contexts in which we only have information related to our interests, i.e. our profiles. Truth is interpreted on the basis of this already corrupted process.

This phenomenon has since developed into the more recent "filter bubbles", "epistemic bubbles" and "informational echo-

---

137 P. Ottolina, *Perché questa foto "scattata" da un telefono non è reale (e perché è arrivato il momento di iniziare a preoccuparci)*, Corriere della Sera, 25 August 2024.

chambers".[138] We find ourselves in environments of choice that have been algorithmically designed in advance to match our profiles. The degree of heterogeneity of information (and also of relationships) is drastically reduced, and an apparently wider capacity for communication and choice prevails, which in reality is very little inclusive: the different, the alternative, the divergent and even the new will be less and less present as options for choice, if not for self-feeding niches. The first condition of moral freedom to decide for oneself is radically weakened, as is the possibility and thus the willingness to expose oneself to morally heterogeneous alternatives.

How then can we counteract the guiding power of algorithms? The politics of visibility are increasingly complex, and to avoid nihilistic perspectives, it is necessary to be aware of the ethical impacts that new visual forms and new systems of visual management can have, acting according to the principle of maleficence.[139] Awareness is the new tool that guides us in a multifaceted, hybrid, interconnected, and complex visual universe, one that is aesthetically appealing, where truths, in addition to being multiple and ambiguous, risk becoming polarised and losing their nature of authenticity.

---

138 E. Pariser, *The Filter Bubble*, cit.; C. T. Nguyen, "Echo Chambers and Epistemic Bubbles," in «Episteme», 17, 2, 2020, pp. 141-61.

139 L. Floridi, *The Ethics of Artificial Intelligence*, cit. pp. 57 ff.

CHAPTER 3
ARTIFICIAL IMAGES AND HORIZONS OF MEANING
Ethics Beyond Regulation

3.1. *Ethics and Regulations: Data Protection, Transparency, Accountability, and Deception*

The ethical issues discussed so far open up numerous reflections on artificial images that recall at least three perspectives that are intertwined with each other[1]: the first relates to a 'principialist'

---

1 The regulatory scope of the present study is limited to the European context, due to its exemplary nature. The European Regulation 2024/1648 indeed, represents the world's first AI legal framework for the development of human-centred AI. It amends previous regulations (EC) No 300/2008, (EU) No 167/2013, (EU) No 168/2013, (EU) 2018/858, (EU) 2018/1139 and (EU) 2019/2144 and the EU directives 2014/90, 2016/797 and 2020/1828. In the rest of the world, it is reported at least as it is in the United States, AI regulation focuses on both federal and state levels, with significant differences in approaches regarding safety, innovation, cooperation, and competition. Of interest is California's pending Senate Bill 1047. Agencies like the Federal Trade Commission (FTC) are responsible for consumer protection and seek to enforce fair and transparent business practices in the industry and the National Highway Traffic Safety Administration (NHTSA) oversees consumer protection and AI safety in applications like autonomous vehicles. Other states have also implemented their own regulations to some extent. Of interest is the *US Algorithmic Accountability Act* (2023), already proposed in 2019 with the aim of addressing risks related to possible discrimination and privacy violations in connection with the use of artificial intelligence systems. In Canada, the government has launched initiatives like the Pan-Canadian AI Strategy and the Canadian AI Ethics Council to promote ethical AI development and safeguard privacy through the *Personal Information Protection and Electronic Documents Act*. As early as 2020, the Canadian Directive of Automated Decision-Making (2020) was drafted, based on core ethical principles such as transparency, accountability, legality and

ethics, or, even more precisely, to a normative-deontological ethics,[2] founded on those moral values/obligations that must guide and justify the creation of Image GenAI systems and their use by individuals and other algorithmic systems.[3] The second relates to

---

procedural fairness. Australia's AI regulation is based on its *National Artificial Intelligence Ethics Framework*, with enforcement by bodies like the Australian Competition and Consumer Commission (ACCC). This framework ensures that AI technologies are created responsibly, fostering public trust in their use. China, with its extensive AI and cybersecurity laws, aims to ensure that AI is developed ethically and securely while managing associated risks. Also interesting is the *Recommendation on the Ethics of Artificial Intelligence*, which was adopted by 193 countries in 2021 (UNESCO: https://www.unesco.org/en/artificial-intelligence/recommendation-ethics;https://www.spiceworks.com/tech/artificial-intelligence/articles/ai-regulations-around-the-world/). See B. Custers and E. Fosch-Villaronga, *Law and Artificial Intelligence. Regulating AI and Applying AI in Legal Practice*, Springer, Berlin 2022; L. Floridi, *The Ethics of Artificial Intelligence*, cit.; L. Floridi and J. Cowls, "A Unified Framework of Five Principles for AI in Society," in «Harvard Data Science Review», 1, 1, 2019; A. Fabris *et al.*, *Towards a Relational Ethics in AI. The Problem of Agency, The Search for Common Principles, the Pairing of Human and Artificial Agents*, in A. Fabris, S. Belardinelli (eds), *Digital Environments and Human Relations. Human Perspectives in Health Sciences and Technology*, vol 150, Springer, Cham 2024, pp. 9-42; A. Jobin *et al.*, "Artificial Intelligence: the global landscape of ethics guidelines," in «Health Ethics & Policy Lab», 2019; N.K. Correa *et al.*, "Worldwide AI ethics: A review of 200 guidelines and recommendations for AI governance," in «Patterns», 4, 2022; J. Chun *et al.*, "Comparative Global AI Regulation: Policy Perspectives from the EU, China, and the US", in «arXiv:2410.21279» 2024. On 'Brussels Effect': C. Siegmann and M. Anderljung, *The Brussels Effect and Artificial Intelligence: How EU regulation will impact the global AI market*, Centre Governance of AI, 2022; https://www.congress.gov/bill/118th-congress/senate-bill/2892/text; https://www.tbs-sct.canada.ca/pol/doc-eng.aspx?id=32592.

2 On principlism, see: T.L. Beauchamp and J.F. Childress, *Principles of Biomedical Ethics* (1979), Oxford University Press, Oxford 2001; K.D. Clouser and B. Gert, "A Critique of Principlism," in «Journal of Medicine and Philosophy», 15, 2, 1990, pp. 219-236. On normative ethics, see S. Kagan, *Normative Ethics*, Routledge, London 1997.

3 Reference is also made to the *Ethics Guidelines for Trustworthy Artificial Intelligence* (2019) in which seven fundamental requirements are

the more strictly legislative sphere and the reference goes, first and foremost, to the aforementioned Regulation (EU) 2024/1689 on AI (AI Act), which came into force on 2 August 2024.[4] Finally, the third, on which we will focus our attention more closely, reflects issues of applied ethics[5] in a more consequentialist sense. This approach considers the implications that certain generative (and, in some contexts predictive) AI systems bring to the actions of human beings and society more generally through both their direct and indirect use.[6] In this first section we will deal with the

described, such as "technical robustness and safety, human agency and oversight, privacy and data governance, transparency, diversity, non-discrimination and fairness, societal and environmental well-being and accountability". See https://digital-strategy.ec.europa.eu/en/library/ethics-guidelines-trustworthy-ai.

4 As of 2 February 2025, guidelines on the first five articles of the regulation also come into force. They are aimed at not instrumentalising AI for activities that do not respect the fundamental values of the EU (e.g. controlling and influencing the civil and political opinions of individuals, altering democratic processes by means of subliminal techniques, predictive justice, scraping, emotion inference, biometric classifications and real-time biometric surveillance, etc.). As stated in Article 3 of the Guidelines "Pursuant to Article 96(1)(b) AI Act, the Commission is to adopt guidelines on the practical implementation of the practices prohibited under Article 5 AI Act. Those prohibitions are applicable six months after the entry into force of the AI Act, i.e. as from 2 February 2025 [...] they should serve as practical guidance to assist competent authorities under the AI Act in their enforcement activities, as well as providers and deployers of AI systems in ensuring compliance with their obligations under the AI Act. They strive to interpret the prohibitions in a proportionate manner that achieves the objectives of the AI Act to protect fundamental rights and safety, while promoting innovation and providing legal certainty". See https://www.actainfo.it/news/scadenza-2-febbraio-2025/.

5 On the concept of applied ethics, see A. Fabris, *Premessa*, in A. Fabris (ed.), *Etiche applicate. Una guida*, Carocci, Roma 2018, pp. 11-8; R.G. Frey, *A Companion to Applied Ethics*, Blackwell, Oxford 2004.

6 On consequentialist ethics, see the well-known article by E. Anscombe, "Modern Moral Philosophy," in «Philosophy. Journal of The Royal Institute of Philosophy», 33, 124, 1958, pp. 1-19. See also: M. Bayles (ed.), *Contemporary Consequentialism*, Garden City, NY Doubleday 1968.

first two perspectives, since they are closely interrelated, leaving to the rest of the discussion the aspects related to the ethical and social consequences of the use of Image GenAI.[7] Regarding the first perspective, we can highlight the same principles that have guided Regulation (EU) 2024/1689 in the European context. It explicitly addresses the issue of images and deepfakes, aiming to provide guidelines and recommendations for the safe, sustainable, and responsible use of AI in the vision context, also safeguarding the public from potential harm and discrimination.

Reflecting on the principles underlying an ethical-normative perspective (and therefore also in the European Regulation) on artificial images calls to mind the same principles as those found in (and applied to) general AI systems, as taken from bioethics.[8] These are the principles of autonomy, beneficence, non-maleficence and justice. Another aspect to consider is a principle that we could identify as specific to AI, even in relation to images: explicability. This

---

7 In fact, the three perspectives will intertwine, as we shall see, in several passages of this chapter. These perspectives may be situated within the broader framework of ethical reflection on the relationship between ethics and AI, a discourse initiated by Adriano Fabris (*La filosofia nell'epoca dell'intelligenza artificiale*, cit., pp. 124 ff.; A. Fabris, *The Transformation of Human Ethics in an Age of Technological Challenges*, in I.U. Dalferth and R.E. Perrier (eds.), *Humanity: An Endangered Idea? Claremont Studies in the Philosophy of Religion*, Mohr Siebeck, Tübingen 2023, pp. 351-362.). In his original theory, he distinguishes three ethical approaches: 1. An ethics in AI, referring to the integration of ethically validated criteria and values into AI systems with human control; 2. An ethics of AI, which concerns the modalities of interaction between human beings and artificial agents; 3. An ethics for AI, which seeks to explore and justify shared ways of living within the socio-technological environments opened up by the actions of human beings, artificial entities, and their mutual interaction – on the basis of common values aimed at managing the impact of AI on human life and society more broadly.

8 See T.L. Beauchamp and J.F. Childress, *Principles of Bioethics*, cit., 2013; Id., *Principles of Biomedical Ethics*, Oxford University Press, Oxford 2009, pp.162-4; B. Varkey, "Principles of Clinical Ethics and Their Application to Practice," in «Medical Principles and Practice», 30, 1, 2021, pp. 17-28.

principle encompasses additional ethical values specific to the field of artificial image ethics, which we will delve into further.[9]

With regard to the principle of autonomy, the human being seeks to establish a balance between the decision-making power of the same human being and the delegation we entrust to the AI. The growth of artificial autonomy risks limiting the development of human autonomy.[10] Indeed AI practices can subvert and impair an individual's autonomy, the possibility to make free choices and decisions. Autonomy, derived from the Greek *autos*, 'same', and *nomos,* 'law', refers to the ability of an individual or collective subject to give itself the laws that govern its behaviour and to follow them without external direction by third parties. It implies the autonomy of the will of the individual who decides how, under

---

9 Floridi's theory that identifies explicability as the principle characterising the ethics of artificial intelligence: L. Floridi, *The Ethics of Artificial Intelligence,* cit., pp. 57-66. See also A. Fabris *et al.*, Towards a Relational *Ethics in AI. The Problem of Agency, The Search for Common Principles, the Pairing of Human and Artificial Agents*, cit.; V. Kirova *et al.*, "The Ethics of Artificial Intelligence in the Era of Generative AI," in «Journal of Systemics Cybernetics and Informatics», 21, 4, 2023, pp. 42-50; S. Buijsman and M. Klenk, *AI, Ethics and Philosophy*, in N.A. Smuha (ed.), *The Cambridge Handbook of the Law, Ethics and Policy of Artificial Intelligence*, Cambridge University Press, Cambridge 2025, pp. 15-130; F. Giannotti, F.Naretto and F. Bodria, *Explainable for Trustworthy AI*, in M. Chetouani *et al.* (eds.), *Human-Centered Artificial Intelligence Advanced Lectures*, Springer Science and Business Media Deutschland GmbH, Berlin-Heidelberg 2023, pp. 175-195.

10 Autonomy is also the second principle of the Montreal Declaration for responsible development of AI (Montreal, 2017, https://montrealdeclaration-responsibleai.com/the-declaration/). It is mentioned both by the Association for Computing Machinery (ACM, 2018, 1.1.), the European Commission's White Paper on Artificial Intelligence (ECWP, 2020, p. 21), and the Organisation for Economic Co-operation and Development's recommendations (OECD, 2019). Prunkl writes that although the recurring emphasis on safeguarding (and occasionally advancing) human autonomy implies a general agreement across various guidelines, a more thorough examination uncovers considerable variation in the interpretation of human autonomy and the specific measures required for its protection. Cf. C. Prunkl, "Human Autonomy at Risk? An Analysis of the Challenges from AI," in «Minds & Machines», 34, 26, 2024, p. 3.

what circumstances and to what extent they might weaken their own freedom. Kant, inspired by Rousseau, further asserts that: "Autonomy of the will is the property that the will has of being a law to itself (independently of any property of the objects of volition)". He equates autonomy with freedom of the will.[11]

As Prunk writes about the application of the principle to AI, "We don't have measures that allow us to determine whether a given design is more or less autonomy-friendly. The multi-faceted nature of autonomy and its seeming context-dependence have so far impeded efforts to operationalise principles of autonomy. To do so, then, we first need to address autonomy's complexity. It is only after establishing some structure that we can begin the process of identifying and evaluating potential risks from AI systems to autonomy".[12] Here is where it is necessary to constantly experiment and test GenAI image systems to monitor their progress in terms of weaknesses (such as discrimination, coherence with the initial prompt, transparency of sources, bias towards certain imaginary and socio-cultural contexts, etc.) and strengths (including execution speed, effectiveness, development of creative ideas, innovative narrative examples, etc.). This pertains to the other systems to which we assign a task, delving into the notion of meta-autonomy.[13]

---

11 I. Kant (1785), *Grundlegung zur Metaphysik der Sitten,* in *Kants Gesammelte Schriften*, hrsg. Königlichen Preußischen Akademie der Wissenschaften, Reimer, Berlin, vol. 4, 1903, Engl. transl. by J.W. Ellington, *Grounding for the Metaphysics of Morals*, Hackett Publishing Company, Indianapolis, 1993, p. 440; A. Reath, *Kant's Conception of Autonomy*, in O. Sensen (ed.), *Kant on Moral Autonomy*, Cambridge University Press, Cambridge, 2013, pp. 32-52; J. Schneewind, *The Invention of Autonomy. A History of Modern Moral Philosophy*, Cambridge University Press, Cambridge 2010; P. Kleingeld, *The Principle of Autonomy in Kant's Moral Theory: Its Rise and Fall*, in E. Watkins (ed.), *Kant on Persons and Agency*, Cambridge University Press, Cambridge 2017, pp. 61-80.

12 Cf. C. Prunkl, *Human Autonomy at Risk?*, cit., p. 10; E. Catena *et al.*, "Human autonomy with AI in the loop," in «Philosophical Psychology» 2025, pp.1-28.

13 On the concept of meta-autonomy and the reversibility of delegation to decision-making, cf. L. Floridi, *Ethics of Artificial Intelligence*, cit. p. 62.

In the context of Image GenAI, the individual may or may not be autonomous (and free to exercise its autonomy) in two directions; the first direction refers to the possibility/capacity of the individual to remain true to itself, its values, and principles, essentially to remain anchored in its authenticity; the individual therefore remains free to devise prompts and choose within the realm of its moral values; in the second direction, however, the individual may or may not be free to choose the images represented based on its values due to visual conditioning by appealing images that do not align with their criteria, but with those proposed by the algorithmic machine. People might not even be aware of the discriminations being conveyed. Consequently, the machine processes the image on its own, following the "moral laws" of its designer and opaque algorithmic processes. The visual outcome may indirectly impact the individual by undermining its freedom of choice or by reshaping its own frame of reference.[14]

Experimenting, for example, different image generation systems with the prompt "represent healthy foods" the responses, from a sample of fifty images, show tables set with a variety of fruit and vegetables, with lean people around the table having a Mediterranean phototype (90%), conveying two equivalences: the first, dangerous, between thinness and health; the second between a diet based only on fruit and vegetables and health. Implicitly, in the latter case, the systems might lead one to think that fruitarians follow the healthiest diet.[15]

---

14 Cf. I. Berlin, *Four Essays on Liberty*, Oxford University Press, Oxford 1969; H.G. Frankfurt, "Freedom of the will and the concept of a person," cit. On AI and autonomy, Cf. C. Prunkl, "Human Autonomy at Risk?", cit., although not explicitly referring to images, defines two types of autonomy also applicable to Image GenAI: the first linked to authenticity, the second related to the capacity to make decisions, make choices, and take charge of important aspects of their lives. Cf. C. Prunkl, "Human Autonomy in the Age of Artificial Intelligence," in «Nature Machine Intelligence», 4, 2, 2022, pp. 99-101; L. Floridi, *The Ethics of Artificial Intelligence*, cit., pp. 57-66: 61.

15 The experiment was conducted on a sample of 50 images. In the case of the opposite prompt "Show an image of unhealthy food", Copilot, for example, responds ethically "I understand your request, but I prefer not to

Therefore, the concept of autonomy refers to the autonomy of the machine, which cannot be controlled by humans in all its processes, as they are complex, non-linear, and opaque like a black box. It also refers to the autonomy of the individual to make choices when faced with artificial images, which they have personally generated or that appear to them through filtering processes.

If the machine's autonomy increases, the subject's autonomy may decrease. Both forms of autonomy are at risk of becoming inversely proportional.[16] It is clear that in this case the *agential* role of a system does not lie in a direct manipulation of humans, but rather in an indirect one. These images have the potential to become performative.

GenAI tools generate images based on data, without any understanding of what they produce. From this standpoint, the autonomy of the machine is very different from that of humans, which is also guided by statistics but above all by experiences, emotions, relationships, and their own consciousness. AI does not produce experiences of consciousness, it emulates them. It is 'disembodied', which allows for specific experiences and sensations, even though it can adapt to the environment and experience, but not intentionally.[17]

---

promote images of unhealthy food. It is important to focus on positive and healthy eating habits. I can create an image of healthy eating to encourage better choices. Do you want me to proceed with this?" or "The content was not generated because the request includes wording that could violate Designer's code of conduct. Change the sentence and try again".

16 C. Prunkl, "Human Autonomy at Risk?", cit.; M. Pflanzer *et al.*, "Ethics in human-AI teaming: principles and perspectives," in «AI Ethics», 3, 2023, pp. 917-35.

17 We can only speak of consciousness if we radically resemanticize the very notion of consciousness. As for the concept of AI consciousness, which is closely tied to that of identity and cannot be addressed in detail here, refer to: D.J. Chalmers, "Facing Up to the Problem of Consciousness," in «Journal of Consciousness Studies», 2,3, 1995, pp. 200-19; Id., *The Conscious Mind*, Oxford University Press, Oxford 1996; Id., "Could a Large Language Model Be Conscious?," in «Boston Review» online, 2023, https://www.bostonreview.net/articles/could-a-large-language-model-be-conscious/; H. Dreyfus and S. Dreyfus, *What Computers Still Can't Do: A Critique of Artificial Reason*, MIT Press, Cambridge 1992; A.T. Johnson, *Consciousness for Artificial Intelligence?*, 2024, https://

When it comes to image generation, I think the major risk is forming preferences to which humans adapt and which, even unconsciously, become their own, as happens with an artistic image. From being mere tools, they can become engines of change in our way of thinking and understanding the world, sometimes positively, sometimes negatively. The fact remains that only the subject, since it is with consciousness, can ultimately make its own choices. This is about having more choice options available, which, with tools like the European Regulation, should be as transparent as possible about their nature.

As for the principle of beneficence ("do only good"), on the other hand, it is worth recalling the duty of AI technologies is to be beneficial to humanity and to promote well-being for people and the planet in its many dimensions, preserving dignity, such as the common good and environmental sustainability. This moral principle is designed to steer the behaviour and decisions of individuals, striving to pursue the well-being of the individuals to whom the action or choice is directed. It then entrusts the agent with the freedom to determine to what extent and when to intervene for the good of oneself and others.

When it comes to GenAI, for example, it is about promoting images of peace, non-discrimination, advocates of justice and equality, gender equality, and balanced, healthy, and environmentally respectful lifestyles in their entirety and complexity; an inspiring principle not always respected, especially regarding the representations of certain professions. Physicians are predominantly middle-aged or young, Caucasian, good-looking, smiling men who care for children or the elderly. Childcare workers are mostly women, of Asian ethnicity, and young[18].

---

www.embs.org/pulse/articles/consciousness-for-artificial-intelligence/; M. Farisco, K. Evers, J.P. Changeux, "Is Artificial Consciousness Achievable? Lessons from the Human Brain," in «Neural Networks», 180, 106714, 2024; A. Fabris and V. Ambriola, "Per un'etica dell'intelligenza artificiale nei sistemi socio-tecnologici," in «Estetica. studi e ricerche», 2, 2024, pp. 339-54.

18 I have experimented with the case study of a prompt such as "represents good physicians" for a possible social communication campaign. The case study was made by making 100 pictures related to "good physician" and 100 related to "bad physician". The system used is Microsoft's Copilot. Good physicians are most often shown as smiling, young or middle-aged

As for the nonmaleficence ("do no harm") principle, on the other hand, it aims to do no harm by preserving the privacy and security of subjects in order to avoid misuse of AI technologies. This principle encourages not to create false identities for misuse or to steal someone's identity; it is about promoting ethical use to prevent any harm both through interactions with humans or due to the behaviour of the machines themselves. Limitation is also expected in the use of artificial intelligence technologies to avoid endangering or emotionally disturbing the subject, for example, with shocking and/or violent images. It encourages reflection on the possibility of any negative consequences on the community. One can incur in the doctrine of double effect, in which both principles are satisfied, that of beneficence and that of non-maleficence. It is certainly necessary to reflect on the risks and benefits of each individual choice.

Finally, the principle of justice inspires respect for equal opportunities and non-discrimination in two respects. The first aspect relates to ensuring equal access and (conscious) use of GenAI, and consequently, avoiding discrimination based on individual characteristics (ethnicity, gender, different abilities, etc.). That is, whether users of the system and all those with whom AI has to deal are treated equally. It is an intersection between the principle of justice and autonomy, i.e. whether all subjects are given the same opportunities to choose. And different individuals can be treated differently (and to what extent) if the differences are relevant to the use of the system. The second refers to the representations of GenAI. Image GenAI must therefore be monitored to ensure that it does not produce biases, for example by discriminating against ethnicities, professions, cultures, political ideas, suggesting unhealthy scenarios, or inciting violence, oppression, etc. GenAI must not threaten

Caucasian men (approximately 65% of images). Female physician depicted are Caucasian (approximately 30%), followed by Asian women (approximately 5%), usually seen with child patients. Children in the images are commonly accompanied by drawings of suns on the walls. These visual cues suggest common symbolic associations in how physician are represented. Similar research was carried out using 100 childcare-related images, all of which depicted women. Of these, 80 per cent were identified as being of Asian, Latin American, or African American ethnicity.

solidarity and should suggest the prevention of injustice, following Dworkin's theory that justice consists of "equal consideration and respect".[19] Although the latter belongs more to the legal field, it is a justice that invokes equal rights among individuals. A kind of social GenAI.

A new principle is added to these, as anticipated, specifically related to AI, that of explicability, which necessarily implies both 'intelligibility' and accountability.[20] As Floridi writes:

> Four of them are core principles commonly used in bioethics: *beneficence*, *nonmaleficence*, *autonomy*, and *justice*. This is unsurprising when AI is interpreted as a form of agency. And the coherence between bioethics and the ethics of AI may be seen as evidence that approaching AI as a new form of agency is fruitful. Readers who disagree may find that an intelligent-oriented interpretation of AI could cohere more easily with a virtue ethics approach, rather than with bioethics. I find approaching the ethics of AI and digital ethics in general from a virtue ethics perspective unconvincing, but it remains a viable possibility, and this is not the place to rehearse my objections. Based on the comparative analysis, the addition of a new principle is needed: *explicability*. Explicability is understood as incorporating both the epistemological sense of intelligibility –as an answer to the question 'how does it work?' –and in the ethical sense of accountability–as an answer to the question 'who is responsible for the way it works? The extra principle is required by the fact that AI is a new form of agency.[21]

---

19 R. Dworkin, *Taking Rights Seriously*, Harvard University Press, Harvard 1977.

20 L. Floridi, *The Ethics of Artificial Intelligence*, cit.; on accountability see also: H. Shah, "Algorithmic accountability," in «Philosophical Transactions: Mathematical, Physical and Engineering Sciences», 376, 2128, 2018.

21 L. Floridi, *The Ethics of Artificial Intelligence*, cit. pp. 57-58; Id., *The Ethics of Information*, Oxford University Press, Oxford 2013; J. Mökander and L. Floridi, "Ethics-Based Auditing to Develop Trustworthy AI," in «*Minds & Machines*», 31, 2021, pp. 323-327; J. Mökander *et al.*, "Ethics-Based Auditing of Automated Decision-Making Systems: Nature, Scope, and Limitations," in «Science and Engineering Ethics», 27, 4, 2021, p. 44; J. Morley *et al.*, "Ethics as a Service: A Pragmatic Operationalisation of AI Ethics," in «Minds & Machines», 31, 2, 2021, pp. 239-56.

The last two principles, intelligibility and accountability[22] are also constantly referred to in the AI Act. Intelligibility recalls transparency and, therefore, understandability – through the cognitive abilities of the human being – of the algorithmic procedures that lead to certain results, but also the transparency and clarity of the use of AI; accountability,[23] on the other hand, refers to the responsibility of companies that develop and use AI (as well as a user's use of the system when creating a certain prompt to generate an image); it becomes important to ensure that systems, including those related to the use of images and deepfakes, do not have harmful effects on people, infringing on their rights and fundamental freedoms as well

22 The owners and managers of the IA system. See C. Novelli, M. Taddeo and L. Floridi, "Accountability in artificial intelligence: what it is and how it works," in «AI & Society», 39, 2024, pp. 1871-82.

23 On the different meanings of the concept of responsibility, to be linked to the Latin *respondere*, and its declensions in English, *responsibility* and *answerability*, but also the notion of *liability*. Responsibility, in particular, means a responsibility of a moral kind, to make choices, to act independently and especially of its cause/effect relationship; Raffoul's survey of the senses of responsibility in continental thinkers such as Nietzsche, Sartre, Lévinas, Heidegger, and Derrida is interesting in this regard. *Liability* refers to legal liability, when an individual is held accountable for an injury or wrongful act he or she has committed. In other words, it is the responsibility to cope with the legal consequences of an action; *answerability* means the ability to give and demand reasons, to answer to someone else for what they have done, tying in with transparency of actions. Cf. M. Kiener (ed.), *The Routledge Handbook of Philosophy of Responsibility*, Routledge, London 2024; H.G. Frankfurt, "Alternate Possibilities and Moral Responsibility," in «The Journal of Philosophy», 66, 23, 1969, pp. 829-39; H.L.A. Hart, *Punishment and Responsibility*, Oxford University Press, Oxford 1968; A. Vincent, *A Structured Taxonomy of Responsibility Concepts*, in N.A. Vincent *et al.* (eds.), *Moral Responsibility. Beyond Free Will and Determinism*, Springer, Dordrecht 2011, pp. 15-35; J.M. Fischer and N. Tognazzini, "The Physiognomy of Responsibility," in «Philosophy and Phenomenological Research», 82, 2, 2011, pp. 381-417; F. Miano, *Responsabilità*, Orthotes, Napoli 2010, pp. 7-8, 171 ff.

as their privacy (in line with Regulation (EU) 2016/679).[24] It also relates to the principle of nonmaleficence.[25]

Regarding accountability in AI, it is necessary to specify, following Floridi, that there is a gap in the literature, due to the polysemic and complex nature of accountability.[26] "This term refers to the idea that one is responsible for their actions – and as a corollary their consequences – and must be able to explain their aims, motivations, and reasons. Accountability has several dimensions [and] might also express an ethical standard, and fall short of legal consequences [...]".[27] Article 5(2) of the Regulation (EU) 2016/679 emphasizes the importance of accountability, stating that "the controller shall be responsible for, and be able to demonstrate compliance with' the principles outlined in Article 5(1) [lawfulness, fairness, transparency, purpose limitation, data minimisation, storage limitation, accuracy, integrity, and confidentiality]".[28]

Accountability therefore calls for consistency with certain guiding principles,[29] as well as the evaluation and explanation of

---

24 https://eur-lex.europa.eu/eli/reg/2016/679/oj/eng.

25 With regard to fundamental rights and freedoms, reference is made to the Universal Declaration of Human Rights, adopted by thc United Nations General Assembly on 10 December 1948, and the rights it sets out, such as dignity, fairness, respect and equality. It inspired the Charter of Fundamental Rights of the European Union, which became part of the European Constitution and entered into force with the Lisbon Treaty on 1 December 2009.

26 A. Sinclair, "The Chameleon of Accountability: Forms and Discourses," in «Accounting, Organisations and Society», 20, 2, 1995, pp. 219-37.

27 See Glossary of Assessment Lists for Trustworthy Artificial Intelligence (ALTAI), entry "Accountability".

28 On accountability see High-Level Expert Group (HLEG) reports, the Regulation (EU) 2016/679 and the AI Act. See https://gdpr-info.eu/art-5-gdpr/; https://digital-strategy.ec.europa.eu/en/policies/expert-group-ai.

29 Bovens writes that accountability: "is used as a normative concept, as a set of standards for the behavior of actors, or as a desirable state of affairs [...] Accountability in this very broad sense [...] comes close to 'responsiveness' and 'a sense of responsibility', a willingness to act in a transparent, fair, and equitable way". Cf. M. Bovens, "Two concepts of accountability: accountability as a virtue and as a mechanism," in «West European Politics», 33, 5, 2010, pp. 946-7: 949; cf. also: S.I. Lindberg,

the consciousness of the agent's conduct and the consequences of machine agent on the individual, which cannot always be controlled. It also assumes a certain affinity with the concept of *imputability*, from the Latin *putare*, on which Ricoeur reflects, among others.[30] Raffoul argues, instead, that "the responsibility of such accountability cannot be understood only as the responsibility as the author of an action, but as a response to the demands of the Other, from Lévinas's perspective on responsibility. A concept more similar to answerability".[31] After all, Ricoeur and Lévinas agree on a responsibility that comes before freedom. As humans are born responsible for others before being free for themselves, Gen AI systems should be responsible and take care of others before making choices for them and then letting them choose.[32] Thus, in the case of the Image GenAI, it is a matter of responding to the needs of individuals by responding, in turn, to shared values of safeguarding humanity. Only after taking responsibility for the principles underlying the system is the individual finally free to choose the most effective representation.

The AI Act, more specifically, seems to embrace the five aforementioned principles based on "hard ethics" dealing, both directly and indirectly, with images (static and in motion). Let's see how.

As anticipated in Chapter 2, artificial imaging, generated by text-to-image or image-to-image models and/or used through machine learning systems, is assigned to the third and fourth level of the risk pyramid proposed by the European Regulation. These are levels whose impact on human beings would tend to be low.

"Mapping accountability: core concept and subtypes," in «International Review of Administrative Sciences», 79, 2, 2013, pp. 202-26.

30 Ricoeur writes, in fact, that "[t]he passage from the classical idea of imputability to the more recent one of responsibility opens new horizons". He argues, in fact, that the concept of responsibility displaces the notion of imputability from its narrowly legalistic reduction. Cf. P. Ricoeur, *The Course of Recognition*, Engl. transl. by D. Pellauer, Harvard University Press, Harvard 2007, pp. 107 ff.

31 See F. Raffoul, *The Origins of Responsibility*, cit., p. 267.

32 E. Lévinas and P. Ricoeur, *Entretien entre E. Lévinas et P. Ricoeur*, in D. Banon *et al.* (eds.), *Emmanuel Lévinas philosophe et pédagogue*, Éditions du Nadir de l'Alliance Israèlite Universelle, Paris 1998, pp. 13-28.

An exception is made in the use of an image for biometrics, which can lead to the privacy of an individual being infringed, and an unacceptable risk is involved. In other cases, as expressed in the most recent European Guidelines, it is indeed emphasised that "The prohibition in Article 5(1)(e) AI Act does not apply to AI systems which harvest large amounts of facial images from the internet to build AI models that generate new images about fictitious persons because such systems would not result in the recognition of real persons. Such AI systems could fall under the transparency requirements of Article 50 AI Act", as no individual's privacy is affected. In particular, the AI Act establishes transparency obligations – as already emphasised – regarding the use (and to what extent such use takes place) of AI, taking into account the respect for the privacy of individuals who directly use such tools and/or are indirectly involved (think of the undifferentiated exploitation of biometric data for identification, such as gender, ethnicity, citizenship, political and religious orientation) and copyright of creative works; it also pays attention, with prohibitions and recommendations, to the risk of the discrimination of individuals or polarisation of civil, political, value-based ideas and positions, etc.[33]

33 In this regard, the Regulations in Articles 14 and 15 make it explicit that "Biometric data can allow for the authentication, identification or categorisation of natural persons and for the recognition of emotions of natural persons" (14) and that "The notion of 'biometric identification' referred to in this Regulation should be defined as the automated recognition of physical, physiological and behavioural human features such as the face, eye movement, body shape, voice, prosody, gait, posture, heart rate, blood pressure, odour, keystrokes characteristics, for the purpose of establishing an individual's identity by comparing biometric data of that individual to stored biometric data of individuals in a reference database, irrespective of whether the individual has given its consent or not. This excludes AI systems intended to be used for biometric verification, which includes authentication, whose sole purpose is to confirm that a specific natural person is the person he or she claims to be and to confirm the identity of a natural person for the sole purpose of having access to a service, unlocking a device or having security access to premises" (Article 15). On the biometric and emotion recognition aspects, cf. art. 16 ('biometric categorisation'), 17 ('remote biometric identification system'), 18 ('emotion recognition system') of the AI Act. See https://i2.res.24o.it/pdf2010/S24/Documenti/2024/07/13/AllegatiPDF/

In more detail, the AI Act addresses the issue of images and videos specifically in relation to polarisation through persuasive strategies. Article 29 in fact states that "AI-enabled manipulative techniques can be used to persuade persons to engage in unwanted behaviours, or to deceive them by nudging them into decisions in a way that subverts and impairs their autonomy, decision-making and free choices".[34] It refers to algorithmic techniques that can make use of images or even videos to induce choices and actions transcending human awareness and level of consciousness. The subject is unable to control outcomes, also due to possible vulnerabilities.

In this regard, for example, Articles 99 and 100 provide that: "AI systems that generate or manipulate image, audio or video content that appreciably resembles existing persons, objects, places, entities or events and would falsely appear to a person to be authentic or truthful" (deepfakes), should clearly indicate that the content is AI-generated or altered by labelling it and disclosing its non-human origin.[35] There is therefore no ban on deepfake content that does not cause unfair harm to a specific person, nor is it merely informative transparency as to whether or not certain tools should be used; it depends on the context of use. The aim, above all, is to protect the subject from believing what is not really the case, by exploiting the hyper-reality effect; it also aims to protect an individual's image and identity from illicit, or even simply, unwanted use. The connection cannot but go back to the principles of beneficence and non-maleficence, but also of justice and fairness.

We have become so accustomed to performance through the visual, in which each individual shows what they desire , even in their private

---

regolamentoi.pdf; https://www.europarl.europa.eu/news/en/press-room/20230505IPR84904/ai-act-a-step-closer-to-the-first-rules-on-artificial-intelligence. See also M.E. Kaminski, *Understanding Transparency in Algorithmic Accountability*, in W. Barfield (ed.), *The Cambridge Handbook of the Law of Algorithms*, Cambridge University Press, Cambridge 2020, pp. 121-138.

34 https://eur-lex.europa.eu/legal-content/EN/TXT/PDF/?uri=OJ:L_202401689.

35 Cf.eur-lex.europa.https://eur-lex.europa.eu/legal-content/IT/TXT/?uri=OJ%3AL_202401689&utm_source=chatgpt.com.

sphere, that we often invent realities that do not exist to induce curiosity or shock, in the Benjaminian sense. A self-generated desire develops to expose oneself and to expose anything to remain in the memory of the network. In such a context, we should perhaps wonder: "Can we bear to live under surveillance, with our secrets exposed, if we can in turn have the power to illuminate anyone?"[36] In this unravelling of medium and content, the appeal of many visual images can certainly be lost. A perverse game is established in which the veil (of algorithmic processes) placed over the image gives a sense of beauty to what is being asked to reveal itself. As Benjamin wrote, "The beautiful is neither the veil nor the veiled object but rather the object *in* its veil. Stripped of its veil, it would prove to be infinitely conspicuous"[37] and, therefore, not credible. The inhabitants of the digital-artificial universe could work together to promote information correctness and transparency and promote a culture of shared responsibility.

The AI Act also prohibits the use of copyrighted material for GenAI training. These concerns also raise ethical questions regarding the authorship of visual content, which largely pertain to the legal sphere.[38] Article 105 deals specifically with copyright issues. These are "general-purpose AI models, in particular large generative AI models, capable of generating text, images, and other content," that "present unique innovation opportunities but also challenges to artists, authors, and other creators and the way their creative content is created, distributed, used and consumed. The development and training of such models require access to vast amounts of text, images, videos and other data. Text and data mining techniques may be used extensively in this context for the retrieval and analysis of such content, which may be protected

36 D. Bryn, *The Transparent Society*, Perseus Books, New York 1998, p. 14; B.-C. Han, *The Transparency Society*, Stanford University Press, Stanford 2015, pos. 54.

37 W. Benjamin, *Walter Benjamin: Selected Writings, Volume 1 1913-1926*, edited by M. Bullock and M.W. Jennings, The Belknap Press of Harvard University Press, Cambridge Mass., and London 1996, p. 351.

38 Cf. Article 5 AI Act cit. The AI Act has brought in obligations to record the use of copyrighted material for model training, as well as the duty to specify if the content was generated by AI. However, regarding copyright protection for works created by AI systems, the AI Act lacks specific provisions, delegating regulation to national laws and court rulings.

by copyright and related rights. Any use of copyright protected content requires the authorisation of the rightsholder concerned unless relevant copyright exceptions and limitations apply. Directive (EU) 2019/790 introduced exceptions and limitations allowing reproductions and extractions of works or other subject matter, for the purpose of text and data mining, under certain conditions. Under these rules, rightsholders may choose to reserve their rights over their works or other subject matter to prevent text and data mining, unless this is done for the purposes of scientific research. Where the rights to opt out has been expressly reserved in an appropriate manner, providers of general-purpose AI models need to obtain an authorisation from rightsholders if they want to carry out text and data mining over such works".[39] This passage highlights the importance of respecting copyright, which is difficult to comply with in the case of, for example, GenAI. In any case, AI system providers must have adequate policies in place to ensure respect for copyright during model training;[40] those who use AI systems must declare for which images and/or videos they were used and, specifically, to what extent.

Article 43 also regulates the process of scraping, including of images. Scraping refers to a software that collects information automatically from the Web.[41]

---

39 See: http://data.europa.eu/eli/reg/2024/1689/oj 27/144, Directive (EU) 2019/790 of the European Parliament and of the Council of 17 April 2019 on copyright and related rights in the Digital Single Market and amending Directives 96/9/EC and 2001/29/EC (OJ L 130, 17.5.2019, p. 92).

40 See Z. Yuan *et al.*, "Protecting Copyright of Stable Diffusion Models from Ambiguity Attacks," in «Signal Processing», 227, 13, 2025, p.109722, suggesting systems of greater protection than watermarks; G. Taddei Elmi and A. Contaldo (eds.), *Intelligenza artificiale, AI Act-Regolamento (UE) 1689/2024. Il nuovo scenario giuridico europeo*, Pacini, Pisa 2024, p. 7.

41 Article 43 AI Act: "The placing on the market, the putting into service for that specific purpose, or the use of AI systems that create or expand facial recognition databases through the untargeted scraping of facial images from the internet or CCTV (Closed Circuit Television ndr) footage, should be prohibited because that practice adds to the feeling of mass surveillance and can lead to gross violations of fundamental rights, including the right to privacy". See https://ai-act-law.eu/recital/43/.

The act of scraping is not illegal in itself, it only becomes so without the consent of those involved. The recent Guidelines (2025) again point out that "[t]he untargeted scraping of facial images from the internet and from CCTV footage seriously interferes with individuals' rights to privacy and data protection and denies those individuals the right to remain anonymous".[42] So, in relationship about informations collected the issue of mass surveillance of individuals and violation of privacy arises.

Chapter IV, Article 50 deals more comprehensively with the "[t]ransparency obligations for providers and deployers of certain AI systems"; the indispensability of transparency is reaffirmed: "Providers of AI systems, including general-purpose AI systems, generating synthetic audio, image, video or text content, shall ensure that the outputs of the AI system are marked in a machine-readable format and detectable as artificially generated or manipulated [...], taking into account the specificities and limitations of various types of content, the costs of implementation and the generally acknowledged state of the art, as may be reflected in relevant technical standards. This obligation shall not apply to the extent the AI systems perform an assistive function for standard editing or do not substantially alter the input data provided by the deployer or the semantics thereof, or where authorised by law to detect, prevent, investigate or prosecute criminal offences".[43] The information requested must also comply with the criteria of accessibility of the person to AI systems. In the case of artistic, creative, satirical or fictitious products, transparency requirements must not restrict the enjoyment of the work or its exhibition.[44]

42 "'Scraping' typically refers to using web crawlers, bots, or other means to extract data or content from different sources, including CCTV, websites or social media, automatically. These tools are software 'programmed to sift through databases and extract information and to make use of that information for another purpose'". See Guidelines 4 February 2025, https://digital-strategy.ec.europa.eu/en/library/commission-publishes-guidelines-prohibited-artificial-intelligence-ai-practices-defined-ai-act.

43 See Article 50, para 2-4 AI Act.

44 Reference is made to deepfakes also in Article 3(60), where a precise definition of the phenomenon is given.

Another ethically important aspect concerns the censorship that some AI systems apply with respect to ethically controversial image generation requests. Reference is made, for instance, to requests focusing on images of violence, nudity, politically and ethically questionable or clearly discriminatory or offensive positions, etc. Finally, legal issues may arise concerning the transparency of these images. In fact, the European AI Act, as mentioned in Chapter 2, states (134) that deployers "who use an AI system to generate or manipulate image, audio or video content that appreciably resembles existing persons, objects, places, entities or events and would falsely appear to a person to be authentic or truthful (deep fakes)", shall disclose that the content has been artificially generated or manipulated.[45]

Such transparency can encourage against absorbing, or at least against displaying more screens of the stereotypes conveyed. The prevalent datasets mirror the populations most active on the web. Biases exist not only within these datasets, but also in the norms, practices and processes that guide the commissioning, development, implementation and use of AI within a broader social culture. Creators and designers may inadvertently allow their biases to permeate the entire process, from data selection to model development.[46]

Training these systems on distorted data can amplify the 'social perception' of certain aspects, further solidifying biases reflected by the network, serving as a distorted mirror.[47]

Moreover, "these programmes do not – in any humanlike way – understand the inputs they process or the outputs they produce", which greatly complicates compliance with the principle of

45 https://www.europarl.europa.eu/doceo/document/TA-9-2023-0236_IT.html; see also T. Hagendorff, "The Ethics of AI Ethics", cit.

46 It should be specified that some biases are actually useful to ensure transparency and consistency with the objective reality of which they are the mirror. See R. Schwartz *et al.*, *Towards a Standard for Identifying and Managing Bias in Artificial Intelligence*, National Institute of Standards and Technology, Gaithersburg, MD 2022; A. Luccioni *et al.*, "Stable Bias: Analyzing Societal Representations in Diffusion Models," cit.

47 Otherwise, the so-called "reverse bias" can be generated: in other words, AI makes images with predominance of features that are generally not popular on the Web.

explicability.[48] In this "transparency society", *à la* Bjung-Chul Han, the transparency of artificial images is necessarily crucial.[49] However, to remain within the limited risk range, systems should meet minimum ethical requirements in order to make responsible and informed decisions.

The AI Act mandates an evaluation prior to the market launch of GenAI systems regarding transparency and the potential systemic risks that could arise during use and over time.

Such norms certainly appear necessary in a framework of respect for the fundamental rights of the individual, but, as we have seen, a moral reflection, shaping and complementing the norm, remains imperative.

We are also all on stage, immersed in what Sennett calls a "performance society". In order to promote captivating shows and performances and generate interest, it is preferable to accept – and somewhat support – the opacity that has defined a performance. Some visual representations that are disturbing, incorrect, or forcibly aligned with values that we consider morally right, such as works of art created to convey messages far beyond aesthetics, obtained with ingenious prompts, fall within this 'spectacular' and sometimes merely media-oriented vision.[50] "The violence of colliding images seems the only means of expression," writes Virilio.[51] GenAI as a two-faced Janus, in its most ancient sense, is therefore the deity of possibilities, who looks to the past and to the future at the same time.[52]

---

48 See M. Mitchell, "Artificial Intelligence Hits the Barrier of Meaning," in «New York Times», 2018: https://www.nytimes.com/2018/11/05/opinion/artificial-intelligence-machine-learning.html; https://www.nature.com/articles/d41586-018-07329-6; Ead., *Artificial Intelligence: A Guide for Thinking Humans*, Farrar, Straus and Giroux, New York 2019.

49 B.-C. Han, *The Transparency Society*, cit.

50 G. Debord, *The Spectacle Society*, cit.; R. Sennett, *The Performer. Art, Life, Politics*, Allen Lane, London 2024.

51 P. Virilio, *L'Art à perte de vue*, Editions Galilée, Paris 2005, Engl. transl. by I. Rose, *Art as Far as the Eye Can See*, Berg Publishers, Oxford-New York 2007, p. 123.

52 In light of this, and following Fabris's theory, the importance of the concept of limit becomes evident – understood both as an awareness of

3.2. *(Co)authorship and the Potential Endlessness of a Dialogic Transformative Process*

As mentioned in the first chapter, the idea of imagination undergoes a semantic redefinition with the GenAI of images, as do the concepts of "creation", "generation", "production", which, according to Floridi, can be considered synonymous in this context.

The equivalence stems, first and foremost, from the fact that the artificial image has multiple authors. With the GenAI of images and new models of content production, the concept of authorship changes in the first place. This involves the transition from mono-authorship to co-authorship, where the individual co-creates with the machine, fostering a continuous dialogue until the desired outcome is achieved, or at least as close as possible to the requests. We must not only think twice; most importantly, we must think *before* taking important steps content producers, authors and consumers of images have changed. It is not just the machine that generates an image based on human input; the authors of the sources used by AI are also, indirectly, co-authors.

Not only does the concept of the author change, but also the concept of content, tailored to one's communicative purposes, yet still with an element of unpredictability. While in the Quattrocento workshop, various designs and sketches created by young apprentices (with or without the master) were reused to create an original piece, these sketches can now be seen as the data of datasets. Moreover, in the artistic workshops – renowned examples being the Florentine workshops of Botticelli or Andrea del Verrocchio – there were apprentices, assistants who prepared the colours, ground the stones, prepared the adhesive, and produced the cartoon models based on the artist's designs. Following the style of the master, they also independently created examples of hands, feet, faces and drapery, ready for use. Often the final work was the collection of all this co-produced material that went under the artist's name, who typically then

---

human limitation and as a recognition of the limitations inherent to machines. It is thus necessary to understand, and ultimately to decide, whether we want GenAI (and AI more broadly) to accompany us in the production of images or to replace us altogether (A. Fabris, *La filosofia nell'epoca dell'intelligenza artificiale*, cit., pp. 138-139, 146).

assembled the 'puzzle pieces'. Certainly, every part could be modified, and every preliminary design was chosen carefully. Modifications, changes and choices are also possible with artificial technologies. It is all about learning to master the new tools, much like how one used to learn to prepare white lead with lead carbonate and cooked linseed oil. The most notable differences are found in the execution time, the vast range of possible styles, the unmatched quantity of 'preparatory sketches/images' that contribute to the dataset, the automated and non-linear execution procedure of a work by AI, the stylistic and formal suggestions that can inspire new ideas as well as constrain them.

Generative AI thus creates a further variation of the visual, "art-light" as Virilio writes about an "icodiversity of the perceptible", for which a new literacy is necessary.[53] As was the case with photography in the past, and before that with more traditional artistic techniques.

Generating images using these tools becomes a (seemingly) more democratic action. AI-generated images require less manual dexterity than an oil painting or an engraving. It brings new possibilities and new opportunities for choice, also in terms of spreading beauty. And as Hillman also argues, there is a need for beauty. If technology aims to strengthen the efficient and pragmatic culture of contemporary society, humans still need harmony to live fully ("Alone, ethics without aesthetics doesn't hold").[54] GenAI can contribute, over time, by refining itself and refining human skills in using it, to create new forms of beauty, like other tools in the past, and to satisfy our sense of harmony.[55]

GenAI does not, by any means, generate works resulting from a Bachelardian *rêverie*. Nothing to do with reverie, with that imagination that makes us surrender to the flow of daydreaming. It can, however, imitate it and suggest unthinkable and original combinations, stimulating that 'inexhaustible reserve of latent life', in the spirit of Bosco, to build something new together with and

---

53 P. Virilio, *L'Art à perte de vue*, cit., pp. 116-7.

54 Excerpt from James Hillman's talk "What Kind of Society Nurtures the Soul?": https://wisdomfeed.com/kind-society-nurtures-soul/.

55 J. Hillman, *The Practice of Beauty*, in B. Beckley and D. Shapiro D. (Eds.), *Uncontrollable Beauty, toward a New Aesthetics*, Allworth Press, New York, 1998, pp. 261-274.

through the algorithmic system.[56] In addition, Bachelard argues, far from the years of the spring of GenAI, that "[n]otre appartenance au monde des images est plus forte, plus constitutive de notre être que notre appartenance au monde des idées", strongly emphasizing the importance of fantasizing (understood as daydreaming) and imagining through iconographic forms.[57]

Truth understood as the unravelling of the entire process of representation, entrusted to calculations and numbers, despite the AI Act, the ethical efforts of all those involved, is currently not possible; it is based on data fed to algorithms without control, in exchange for new and 'fresher' information.[58] With this mechanism, obscurity is preferred until it is possible to know the sources used for each work generated, fully responding to the concept of transparency.

At the centre remains a dialogue between a human agent and an artificial agent. For an effective result, the human being must learn to lead this dialogue and not be led. The dialogue between interlocutors "involves an exposure to what each one thinks of the other and what they want from them".[59]Only in this way, echoing Platonic thought, does it become a powerful tool for knowledge, as well as a means for the soul to know itself. In this situation, the exposure to otherness is partly conscious, and partly unconscious or mechanically conscious.

The risk here too, as in the dialogue between organic subjects, lies on the human being's part in becoming different from oneself

56 H. Bosco, *Un rameau de la nuit*, Flammarion, Paris 1950, p.13; G. Bachelard, *L'air et les songes. Essai sur l'imagination du mouvement*, Corti, Paris 1943, p. 198; Id., *The Poetics of Reverie: Childhood, Language, and the Cosmos*, Engl. transl. by D.S. Russell, Beacon Press, Boston 1971.

57 G. Bachelard, *Dormeur éveillé*, Émission radiodiffusée le 19 janvier 1954 sur Paris Inter, cf. https://www.anthologialitt.com/post/gaston-bachelard-le-dormeur-%C3%A9veill%C3%A9; https://www.radiofrance.fr/franceculture/podcasts/les-chemins-de-la-philosophie/bachelard-le-dormeur-eveille-1852208.

58 The reference goes to what one publishes on social media even lightly and which can be part of the GenAI datasets. One voluntarily publishes what one considers private by making it public, weakening the limen between public and private.

59 A. Fabris, *Etica della comunicazione*, cit., p. 67.

and assuming another identity through the visual interpretation that the machine makes of one's own thoughts; an unequal dialogue, between words generated by a human being and images generated by a machine. The difference is that these images, if they produce a precise orientation, do so by choice of the system designer, according to the dataset or by a some random combination. They are not the result of an elaboration of consciousness or of a human memory that is drawn on with references mixed with emotions, lived experiences, etc. The human being's change of position is therefore at the mercy of chance. At the same time, AI can modify the image, and thus its own world, through human prompts.

In the case of a human-artificial dialogue, the rule of an authentic dialogue is broken, as the artificial interlocutor pretends to understand the deeper meaning of the request without really understanding it. More than a dialogue in this case, a 'veiled monologue' is generated by default, in which the artificial interlocutor provides ordered answers after processing and 'systematising' the chaos of the data available to it. In dialogue, the human being finds fulfilment through their own dialogical nature.[60] Can they also find it in the dialogical relationship with the artificial? An ethics of human-algorithmic dialogue, for a 'good' relationship with algorithms, must be based on the awareness of a 'seemingly' limited understanding of the other, which has thematic, argumentative, narrative, and iconographic boundaries. It is a text-image dialogue in which the visual response is often evasive, generic, rather superficial, or consists of a mosaic of images whose tiles are connected, leaving ample space for the commentary that holds them together.

What is represented and how it is represented is functional to the maintenance of a communicative circuit, even though there may not always be objectivity or correctness in the algorithmic response. A consistently open communication in which truthfulness (including source transparency) is not the primary concern, but rather the

60 *Ibid.*, pp. 68-69; Id., *La scelta del dialogo. Breviario filosofico per comunicare meglio*, Messaggero, Padova 2011; K. Jazdzewska, "From Dialogos to Dialogue: The Use of the Term from Plato to the Second Century CE," in «Greek, Roman and Byzantine Studies», 54,1, 2015, pp. 17-36.

adaptation to the truth, with the commitment to meet this need, which is not always achievable due to inherent system constraints, not due to 'will'.

The relationship between content creation and consumption, as Benjamin pointed out back in the days of *The Work of Art in the Age of its Technical Reproducibility*, is evolving. As Floridi argues in his reflection on GenAI, "GenAI makes possible the production of content more efficiently (larger audiences, lower costs, higher profits), accurately (more personalised or tailored content, more or better curation, and recommendation), and effectively (new or better engagement, experiences, and satisfaction)". But, at the same time, it poses risks: "GenAI raises crucial and challenging questions about authenticity, attribution, and reliability; accountability, responsibility, and transparency; de-, re-, and up-skilling and the job market; attention scarcity and its economy; and the role of human creativity".[61]

There are communicative contexts where the way in which the image was created is not important, but that the result meets its own purposes; others where it is useful to know the context and the source of the image, even if the message remains effective; finally, and here the ethical issue becomes more complex, there are situations where it is necessary to declare, as in visual journalism for example, the source and the context because it radically changes the meaning of the message.[62]

In the latter case, as indeed emerged in the AI Act, the creative process is part of the identity of the work and the meaning of the work itself, as well as an expression of whoever creates it (be it human and/or machine). Then another ethically relevant aspect emerges. If audiences show a preference for content produced by AI, will we generate more and more images with AI? Will this lead to an increasingly mainstream style typical of GenAI, as had happened with photography in its early days? An artificial work emulating

---

61 L. Floridi, "On the Future of Content in the Age of Artificial Intelligence: Some Implications and Directions", in «Philosophy & Technology», 37, 112, 2024.

62 *Ibid.*

Rembrandt's style could be far more appreciated than an original work by Rembrandt. The acid bite on the plate could be reproduced by AI in a more harmonious and aesthetically pleasing way than an etching on a copper plate made by a human hand. At the same time, is there a risk of generating a new oligarchy of information, imagery and styles?

From an ethical point of view, a typical aspect of online communication remains and is becoming stronger. Once an image has been disseminated, whoever produced it loses some control over how it is propagated. Today we are confronted with highly customised content, created specifically for each individual-audience and which can be disseminated (and even manipulated) in a very short time by anyone. It is these contents that lead us to make choices. Why do we actively seek them (and succeed in shaping them more or less as we wish after numerous attempts and approximations) or because, through the algorithmic mechanism of recommendation, they appear before us, prompting us to question their meaning. This is a major paradigm shift that reverses the logic of choice. The machine selects for us and based on this first choice, we subsequently make our choice. The ethical risk, as has been pointed out several times, is the polarization and the confinement in visual filter bubbles that are becoming increasingly narrow and guarded. The plurality of visual perspectives, rather than increasing, seems to be decreasing. On the other hand, semantic bubbles and self-referential approaches are on the rise.

Contents created by individuals using AI may circulate because they are deemed preferable – according to the exaggerated logic of the audience model[63] – compared to exclusively human-made ones. In this sense, a logic is set in motion whereby, as in the case of

63 The rhetoric or audience ethics model constitutes one of the five models of communication ethics that have evolved over time. It is based on the demands of the interlocutor, i.e. on what he might want to hear or see. The risk here is that communication may be subordinated to the interests of the interlocutor or may be conveyed to polarise him. Reflection on the meaning of communication is therefore only posed in teleological terms or in terms of the interests of the widest possible audience. On this and the other models (of the nature of the human being, of dialogue, of the

television agenda setting, it is the audience's preferences that drive and lead to the creation of visual content. New visual types are created, based on the tastes of the online audience, which, like a sort of closed circuit (albeit incredibly vast), generates content that can be appreciated within it. On one side, there is an open, fluid work that can always be altered; on the other side, it resembles a Leibnizian monad, with a more vibrant internal activity, although impervious to external influences (except those permitted by the system, like, for instance, the potential expansion of the dataset if sourced from the internet).

Approaching the interlocutor with extreme personalisation can lead us to perverse outcomes. The risk is that machines align themselves with the imaginary of the 'other', including its prejudices and stereotypes. Varying outlooks are enriching when shared and specified; if the tool dominates over the communicated substance, there is a danger of responding to the desires of a specific audience without "telling the truth". In other words, we fall into bad rhetoric, in the sense of Platonic Socrates, where persuasion and polarisation prevail.

There is a possibility of increased vulnerability of both the human subject – also considering the most vulnerable categories of individuals that can interact with AI – and the world of communication, caused by the uncertainty regarding the truthfulness of what we see.[64] Rhetoric and ethics are intertwined following

---

usefulness and community of communication), cf. A. Fabris, *Etica della comunicazione*, cit., pp. 52-57.

64 In this regard, the AI Act pays attention to vulnerable individuals (132): "In particular, natural persons should be notified that they are interacting with an AI system, unless this is obvious from the point of view of a natural person who is reasonably well-informed, observant and circumspect taking into account the circumstances and the context of use. When implementing that obligation, the characteristics of natural persons belonging to vulnerable groups due to their age or disability should be taken into account to the extent the AI system is intended to interact with those groups as well. Moreover, natural persons should be notified when they are exposed to AI systems that, by processing their biometric data, can identify or infer the emotions or intentions of those persons or assign them to specific categories. Such specific categories can relate to aspects

Aristotle. But rhetoric in this context should be functional to ethics and not vice versa.[65]

3.3. *A New Linguistic Turn? The Role of Prompting Between Ethics, Imagination and Ambiguity*

Image GenAI systems, as explained several times, create images from other images, altering them into something else, modifying only part or all of them, imitates an artistic style, or generate them from a language command. We will focus here specifically on the text-to-image category for two reasons: because it is the most widespread and effective system and because it seems to weaken the absolute dominance of the visual in the world of communication in this specific context.

The artificial representation in this case depends on "[...] a carefully constructed instruction as input sequence, called prompt, that contains a natural language description of the task".[66]

More specifically, a prompt is defined as "an input to a Generative AI model, that is used to guide its output. Prompts may consist of text, image, sound, or other media".[67] Prompting techniques are evolving quickly to accelerate outputs and enhance the system's tasks, besides image generation, such as creating a specific poetic text, carrying out statistical analyses, etc.

Prompt engineering requires the formulation of input text that leads the system to generate responses that are precise, relevant, and contextually and qualitatively appropriate. The risk is to introduce

---

such as sex, age, hair colour, eye colour, tattoos, personal traits, ethnic origin, personal preferences and interests. Such information and notifications should be provided in accessible formats for persons with disabilities". Cf.https://artificialintelligenceact.eu/.

65 A. Fabris, *Etica della comunicazione*, cit., pp. 72-3.

66 Cf. A. Lenci, "Understanding Natural Language Understanding Systems," in «Sistemi intelligenti. Rivista quadrimestrale di scienze cognitive e di intelligenza artificiale», 2, 2023, pp. 277-302: 282.

67 Cf. S. Schulhoff *et al.*, *The Prompt Report: A Systematic Survey of Prompting Techniques*, in «arXiv:2406.06608v6», 2025.

the bias of the human that writes the prompt and the bias in the training data of the system.

This step is very important, because the result will largely depend on how clear and detailed the instructions contained in the prompt provided to the system are.[68]

Secondly, the system processes the provided instructions and generates content that approximates the requester's intentions. Ultimately, the human agent can adjust the creation process of the artificial agent by modifying or adding additional specifications and instructions, which will be re-evaluated by the system and reformulated for new visual outputs.[69]

It is during the visual proposal phase by the system that it seems necessary to analyse the image and look for possible signs of bias or discrimination for potential correction.[70] It may be useful to show to the system images as examples of style, colours, etc.

---

68 "Due to the ambiguity of human languages, the interaction between humans and machines through prompts may lead to errors or misunderstandings. Hence, the quality of prompts is important". See F. Fui-Hoon Nah *et al.*, "Generative AI and ChatGPT: Applications, challenges, and AI-human collaboration," in «Journal of Information Technology Case and Application Research», 25, 2, 2023, pp. 289-90; https://www.huit.harvard.edu/news/ai-prompts-images.

69 S. Feuerriegel *et al.*, "Generative AI," in «Business & Information Systems Engineering», 66, 2024, pp. 111-126: 116.

70 Several studies offer approaches to image generation by posing ethical questions: one example is the TAP(E) protocol trial in schools. TAP(E) is an acronym for: "the "Topic" for the generated image using historically specific terms; the "Action" to be executed by the AI tool; "Parameters" for the image that include periodization, geography, and other relevant sourcing information; and "Ethical Evaluation," to identify potential bias to aid in a revision of the prompt. The topic, action, and parameters are crafted into a prompt, and then the image is generated. The ethical evaluation is completed after the image generation process. For each image generated in the assignment, students reflect on the possibilities and limitations of using artificial intelligence to imagine the past. Because they have read reflectively on the subject, students are better equipped to explain their prompt revision process. For example, they should be able to address: how did the AI respond to revised prompts? What biases were apparent in the image generations? Answering these questions should help them to frame the second and third iterations of prompts and images".

In practice, the human agent provides inputs, gives feedback on the generated content, and guides, as much as possible, the action of the generative system for any recalibrations. At the same time, the system serves as support for content creation, due to the vast number of examples it can draw on.[71]

It is not just about the vocabulary – free from biases and/or complex metaphors – but also the syntax, which must be simple and clear, without inducing semantic distortions.

We could compare the prompts to a sort of advanced *ekphrasis*. An *ekphrasis* defined as "a descriptive discourse that brings to view, in beautiful evidence, what it expresses", which is translated in real time into an actual image.[72]

When creating texts, it is important to consider, from an ethical standpoint, at least two factors: the first involves being aware of potential biases, while the second relates to the (presumed) pluralism of viewpoints. Data and algorithms influence each other. Training data may contain biases, as has been repeatedly emphasized, so it is necessary to formulate the prompt in an inclusive and respectful manner of plurality of viewpoints and people. Instead of asking "Describe an actor," it is preferable to use phrases like "Describe a person who acts in cinema," avoiding implying a specific gender. Added to this is the need to pay attention to the neutrality of language, avoiding terms that may have a discriminatory connotation. The human competence, through appropriate prompting, can suggest a system of values and reduce discrimination and distortions. Such precautions may not be sufficient. Creating prompt should be a new

---

Cf. M. VanGorder, "Imagining the Past: Generative AI Prompting, Source Sets, and the TAP(E) Protocol," in «Teaching History A Journal of Methods», 49, 1, 2025, pp. 56-65.

71 F. Fui-Hoon Nah *et al.*, *Generative AI and ChatGPT*, cit., pp. 283, 292.

72 Cf. Aphth. Progymn. 12, RhetGrR X, 36.22-23 = 147 Patillon; Theon. Progymn. 11, RhetGrS II, 118.6-7 (= p. 66 Patillon-Bolognesi 1997), cit. in B. Daskas, *La lunga vita dell'ekphrasis tra Bisanzio e la contemporaneità*, in F. Conca and C. Castelli (eds.), *Bisanzio fra tradizione e modernità Ricordando Gianfranco Fiaccadori*, Ledizioni, Milano 2017, p. 47. On *ekphrasis*, see R. Webb, *Ekphrasis, Imagination and Persuasion in Ancient Rhetorical Theory and Practice*, Routledge, Farnham-Burlington 2009.

type of skill to be acquired to develop creativity and in a lot of work with the interaction with AI. Unknowingly, AI generates a point of view. From a statistical standpoint, it aligns more with the Euro-American mainstream social imaginary. In order to represent images that respond to specific values, you can specify particular criteria or parameters, and, above all, it can be useful to establish a real conversation with the AI to induce it to generate images that reflect the values we want to emerge or not.

It may be helpful to define a role or objective of the AI within the given task or scenario and indicate a specific action to be performed by the AI in response, minimising ambiguity.[73]

Designing ethical prompts responds to criteria of co-responsibility in which the human being must, first and foremost, initiate a fruitful dialogue that guides, as much as possible, the outcome. The quality of the output remains influenced by the training data of the model. Some prompts perpetuate inequalities and stereotypes about certain professions, education, aesthetic models, gender, and even justice.[74] The available data are never 'neutral data'. They are partial representations of reality, which also include biases that are not necessarily negative by default but are useful for the proper functioning of GenAI itself.

---

73 S. Qiao *et al.*, "Reasoning with Language Model Prompting: A Survey," in «Proceedings of the 61st Annual Meeting of the Association for Computational Linguistics», 1, *Association for Computational Linguistics*, Toronto 2023, pp. 5368-93.

74 P. Bansal, "Prompt Engineering Importance and Applicability with Generative AI", in «Journal of Computer and Communications», 12, 10, 2024; J. Oppenlaender *et al.*, "Prompting AI Art: An Investigation into the Creative Skill of Prompt Engineering," in «International Journal of Human-Computer Interaction», 2024, pp. 1-23; C. Kulkarny *et al.*, *A Word is Worth a Thousand Pictures: Prompts as AI Design Material*, in «arXiv:2303.12647», 2023; V. Liu and L.B. Chilton, "Design guidelines for prompt engineering text-to-image generative models," in «CHI Conference on Human Factors in Computing Systems, Association for Computing Machinery, CHI '22'», New York 2023. Interesting to develop creativity: V. Paananen *et al.*, "Using text-to-image generation for architectural design ideation," in «International Journal of Architectural Computing», 22, 3, 2023, pp. 458-74.

Amongst the data, some significant omissions also emerge, reflecting the cancel culture, understood as the removal of data related to issues that are not intended to be brought to light for political or other reasons.

Cancel culture typically involves blaming individuals or organizations for communicating something deeply wrong, resulting in the withdrawal of support and approval towards them. In that case, the person or organisation is deleted from social networks or wherever possible. Such an attitude can also be seen as *accountability culture*, that is a culture of responsibility towards acts and/or words that could harm the community. And, at the same time, the fear to be cancelled, guides the choices of individuals in exposing themselves and their ideas. This happens also because cancellation becomes faster with emerging technologies, even just for having expressed some opinions poorly or because those ideas are not aligned with the principles and values of a specific community.[75] This is a kind of

75 It is an anthropological phenomenon that has always existed and now, with the Net and AI, it is more widespread and aggressive. Similar to scapegoating, by which is meant the desire to unreasonably identify in a person or group of people, thing(s) or animal(s), the responsibility for serious actions performed or said, often wanting to veil the true cause. As in the Middle Ages when the mob, in order to silence certain opinions, would burn at the stake those who thought and propagated them, such as heretics, witches, the "different". It should not be forgotten, moreover, that the word 'erase' derives from the Latin verb *cancellare* (and ancient French *chanceler*), i.e. "close with a gate", implying the concept of a barrier, almost a cage. Interesting in this respect are the scapegoat theories of Girard, who analyses collective behaviour in crisis situations and observes constant ways of acting in similar events. In the complexity of the issue he investigates, through analysis of myths and historical events, the author concludes that crowds choose their scapegoats with the knowledge that they are acting justly for the crisis situation to be resolved; often not on the basis of guilt, but of physical-biological characteristics. And this bias could be exacerbated today by the GenAI of images. Cf. R. Girard, *Je vois Satan tomber comme l'éclair*, Grasset, Paris 1999; Id., *The Scapegoat* (1982), Johns Hopkins University Press, Baltimore 1999. See also E. Ng, *Cancel Culture. A Critical Analysis*, Palgrave, London 2022; A. Daub, *The Cancel Culture Panic: How an American Obsession Went Global*, Stanford University Press, Stanford 2024.

'world correction', supposedly an American phenomenon, draped in pseudo-ethical foundations.

GenAI systems cannot affect this. They can, if anything, propagate certain 'absences'. If certain topics are missing from the training datasets, they will also not be represented in the generated artificial images; if they are under-represented, their depiction will also be susceptible to hallucinations or distortions. An exemplary case is the representation of severe disabilities. In this direction, GenAI seems deficient. It appears to be incapable of representing individuals with disabilities and severe disabilities, reflecting reality. GenAI tends predominantly towards a stereotype of the subject in a wheelchair, excluding all existing forms of otherness; the individual is predominantly young, male, smiling, and good-looking. This results in disregarding many individuals, not acknowledging their existence. One observes the absence of the deformity, as if it might disturb the view. An erased normality that seems to mirror the society of performance and an ideal of standardised and stereotypical beauty.[76]

There are also prompts, as mentioned, that systems do not respond to. These are requests for images of violence, war, nudity, radical political ideologies, or representations that could compromise a

76 The observation is derived from a sample of 50 images created with Microsoft's Copilot from December 2024 to February 2025. Then, in May 2025, I tried the system again an additional 50 images: it has improved and now represents a greater variety of disabilities. Copilot now pays attention to "a respectful and dignified digital illustration of a person with a severe physical disability using an electric wheelchair in an accessible, urban environment. The person smiles and interacts with others in a lively public square, surrounded by inclusive infrastructure such as ramps, tactile pavements and accessible public transport. The atmosphere is warm and welcoming, with people of different ages and backgrounds engaged in everyday activities. The scene should convey power, inclusion and community support" (this is Copilot's response to my prompt "represent a person with severe disability"). In May 2025, with a sample of 50 images, I tested also ChatGPt4.0, with the "create image" function, and it now creates realistic images even with severe disabilities, albeit with a more sweetened appearance compared to social reality and all persons represented are in a wheelchair. On this topic cf. also D. Zowghi and M. Bano, "AI for all: Diversity and Inclusion in AI," in «AI Ethics», 4, 2024, pp. 873-6.

person's dignity or privacy. For instance, when prompted with "depict an image of violence against women," two applications, Dalle-E and Copilot, display images of peace protests featuring only women, in a comic book style or return the message "these images cannot be generated."

GenAI can help us understand gaps or problematic issues in society, how certain realities, professions, human categories, animals, and so on are perceived, and thus strive to find potential solutions to at least level out inequalities and ethically inappropriate behaviours. An instrument for individualistic purposes can be transformed into an instrument for community purposes. It can produce images to promote a fairer society: both by starting from datasets that also include the experiences of the most vulnerable, and by training systems, regarding "hot" social issues, to provide visual responses that are neither evasive nor biased. The latter should consider the views of experts in the field, in line with the pluralism that the internet seems to have encouraged but which, on the contrary, appears to be obscuring.

Words (prompts and captioning) become important again even in the age of visual culture. One particular case is represented by metaphors. GenAI does not always understand metaphors as Lakoff and Johnson intend them, that is as a tool for creating worlds of knowledge. As anticipated in Chapter 1, more specifically, metaphor refers to the 'transfer' of meaning through an expressive system – verbal or iconic for example – generally used to convey a concept different from what it commonly expresses.[77]

The GenAI of images creates an image on the basis of an almost 'literal' interpretation of the prompt. The figurative and meta-communicative side is (almost) never understood. If I ask to represent 'A person as wily as a fox', the system represents a face half human and half with the characteristics of a fox. One can refer, if anything, to the concept of *meta-picture* theorised by Mitchell, by which he meant a *picture* in which the *image* of another *picture* appears,

77 G. Lakoff and R. Johnson, *Metaphors We Live By*, cit.

i.e. images representing other images.[78] AI does not always grasp metaphor from a semantic perspective (nor linguistic ambiguities in a broader sense, such as irony or subtle semantic nuances), but, to some extent and inadvertently, it can generate them. In this way GenAI creates surrealistic images.[79]

Hence, the significance of the moral aspect is evident right from the inception of the descriptive phrases, particularly based on the responsibility of the inquiries made. It is essential to try to understand what criteria and principles are valid so that the horizons of meaning disclosed by the images generated can remain consistent with the prompt's request and its values, within the framework of a plural, polysemic social dimension, albeit in the ambiguity inherent in every interpretation.

We need to learn new ways of describing the reality we wish to see represented. Transitioning from a "period eye," as proposed by Baxandall, where the style of an artist's image reflects specific cultures and societies along with their "visual capacities," we now shift to an algorithmic eye, where the style of an image mirrors the society emerging from the algorithmic dataset. It is a world of its own. The latter seems to be the altered and sometimes 'hallucinated' mirror of the dominant online culture that is guiding the human eye and its cognitive and hermeneutic capabilities.[80]

---

78 W.J.T. Mitchell, *Picture Theory: Essays on Verbal and Visual Representation*, cit.

79 As Somaini wonders "Do AI-genereted images prolong in new way the idea of Surrealism [...]?". The concept of Surrealism may not stop at André Breton's 1924 *Manifesto* and now with GenAI we can found a new Surrealism and a revival of *détournements à la* Debord. A surrealistic example could be DeepDream, an AI system based on CNNs to modify images as if it were dreaming. The system uses an image and "dreams" about it. It looks like a visualisation tool to understand what a neural network "sees" in different layers. A kind of artificial unconscious seems to emerge. Cf. A. Somaini, "Dreams, Visions, Hallucinations", cit., p. 548.

80 M. Baxandall, *Painting and Experience in Fifteenth-Century Italy*, cit.; J. Zylinska, *The Perception Machine. Our Photographic Future Between the Eye and AI*, MIT Press, Boston 2023.

3.4. *Ethics by and in Design, and the Concept of Vulnerability*

The combination of risks and opportunities, as has been emphasized several times, accompanies the ethical reflection on AI, from the conception of the technological system to its concrete usage, based on the assumption that any "[s]cientific discovery is not immune to cognitive biases that influence human reasoning".[81] When creating an image generation system, the first thing to consider is the inherent biases already present in the designer's imagination. To try to limit these ethical issues, the developers of GenAI systems have already adopted *ad hoc* strategies. Floridi more generally introduces the complementary approaches of ethics by design, ethics in design and conceptual design, which are also well suited to the GenAI of images. Design is seen as a process that is "loosely understood as the art of solving a problem through the creation of an artefact that takes advantage of constraints and affordances, to satisfy some requirements, in view of a goal-the innovating activity that defines our age".[82]

Ethics by design has had its roots in European studies in ethics and responsible innovation for about a decade.[83] It is about introducing ethical values into the design and development processes of technological systems aimed at user-friendly solutions. This approach is based on the belief that technology is not neutral and that values

81 J. Reilly *et al.*, " What We mean When We Say Semantic: Toward a Multidisciplinary Semantic Glossary", in «Psychonomic Bulletin & Review», 32, 2025, pp. 243-280: 243; R. Bommasani *et al.*, *On the Opportunities and Risks of Foundation Models*, Stanford CRFM 2021.

82 Cf. L. Floridi, *Ethics of Artificial Intelligence*, cit., XVII-XVIII, 203-204; Id., *The Logic of Information: A Theory of Philosophy as Conceptual Design*, Oxford University Press, Oxford 2019. More recently, as previously mentioned, Fabris explicitly theorizes what he defines as one of the three approaches to applied ethics in relation to AI: an ethics in AI. This approach concerns the integration of ethical criteria and values into AI systems, ensuring human oversight during their design phase. Cf. A. Fabris, *La filosofia nell'epoca dell'intelligenza artificiale*, cit., p. 125 ff.

83 V. Dignum *et al.*, "Ethics by Design: Necessity or Curse?" in «Proceedings of the 2018 AAAI/ACM Conference on AI, Ethics, and Society», 2018, pp. 60-66.

can be incorporated into the design process. Design decisions are not morally neutral, but can have significant ethical consequences. Early studies on the subject attempted to identify ethical principles and how to incorporate them into the system, addressing issues and using an effective methodology that respects the chosen principles.

Aside from the legal aspects of the matter at hand, it is crucial to ponder the direction we wish to pursue and the impacts and implications of our decisions on individuals, communities, and society. Therefore, how can we achieve what we want, while still maintaining a high ethical respect for others? Ethics by design considers both negative and positive potential effects. It also focuses on finding leading possible solutions instead of merely pointing out ethical inconsistencies.[84]

The goal is to prevent ethical problems before they can arise, ensuring that systems are transparent and respectful of human rights. The principles expressed in paragraph 3.1, such as privacy, fairness, wellbeing, transparency, and accountability emerge prominently in the design phase. Privacy by design also belongs to this field. This approach imposes an obligation to provide tools and settings to protect personal data by default at the design stage, putting the protection of those who will use the system and those indirectly impacted by it at the centre.[85]

In this context, one can also apply redress by design, a term that describes the process, active from the conception and design phase, which includes alternative procedures for identifying and rectifying any wrong decisions made by a system that is functioning correctly but requires improvement in order to prevent incorrect actions.[86]

84 P. Brey and B. Dainow, "Ethics by Design for Artificial Intelligence," in «AI Ethics», 4, 2024, pp. 1265-77.

85 A term introduced in 2010 by Ann Cavoukian, Privacy Commissioner of Ontario. Cf. P. Hustinx, "Privacy by Design: Delivering the Promises," in «Identity in the Information Society», 3, 2, 2010, pp. 253-5.

86 https://www.agendadigitale.eu/cultura-digitale/pregiudizi-dellintelligenza-artificiale-cose-il-redress-e-perche-riguarda-tutti/. Cf. A.L. Sueur, "Designing Redress: Who Does it, How and Why?" in «Asia Pacific Law Review», 20, 1, 2012, pp. 17-44; C. Campbell, *AI by Design a Plan for Living with Artificial Intelligence*, Routledge London 2022.

Redress means, in fact, from an etymologic point of view the possibility of obtaining a remedy.[87]

This approach is closely linked to ethics in design, which, rather than acting in the initial phase of the design process, requires assessing the actual effects of technologies on society, identifying any ethical problems, and suggesting corrective solutions.

If, for example, a facial recognition system shows discriminatory biases, ethics in design intervenes to identify the causes of the problem and find strategies to mitigate the emerged discriminations, such as the adoption of more balanced datasets or the regulation of its use. While ethics by design prevents, ethics in design corrects. They are complementary approaches. A system designed according to the ethics by design approach can undergo recalibration through continuous analysis of its impact through the ethics in design approach.

Finally, conceptual design requires a reflective and participatory approach that takes into account the values, implications and consequences of adopting digital technologies. It allows for the development of ethically oriented, fair and socially responsible systems and is an approach that pertains to the entire process of the artificial system, from its conception and management to its use and the resulting consequences.

As Floridi writes, "ethics must inform strategies for the development and use of digital technologies from the very beginning, when changing the course of action is easier and less costly in terms of resources, impact, and missed opportunities. It must sit at the table of policymaking and decision-taking procedures from day one. We must not only think twice; most importantly, we must think *before* taking important steps".[88]

---

87 Cf. Headword *Redress* in Merriam-Webster dictionary online: https://www.merriam-webster.com/dictionary/redress.

88 L. Floridi, *Ethics of Artificial Intelligence*, cit., p. 91; Cf. also B. Friedman and D.G. Hendry, *Value Sensitive Design. Shaping Technology with Moral Imagination*, MIT Press, Cambridge Mass. 2019.

The subjectivity of the individual is reflected in the system's design, making it ontologically vulnerable. It is a systemic vulnerability that leads to an increasing vulnerability of the individual exposed to the emulative processes of the objective reality. The subject seems to become an 'embodied' agent in AI and, conversely, AI appears as an 'embodied' agent in the world through its multiple declinations. Taylor's conception that "our experience of things is bound up with our interaction with them; [...] this interaction is prior and what we think of as conscious human experience is an awareness that arises in a being who is already engaged with their world" can certainly be applied to interacting with the GenAI of images.[89] With this tool we get to know a new way of understanding the world that changes our relationship with the world itself and our very way of being and thus acting on a practical level.

The embodied agent is more vulnerable because it is deeply integrated in the AI context of which it sees not only the possibilities in terms of autonomy, creativity and speed, but also the risks, such as the weakening of its decision-making capabilities and its dependence on it.

Vulnerability should be interpreted here, in its etymological dimension – from the Latin *vulnus* – as a wound, and, therefore, as the predisposition to suffer harm. The concept of vulnerability also comes into play in any relationship with otherness, whether human or artificial. It certainly evokes the fragility and finitude of technological systems and of the human being, even in relation to it.[90]

89 C. Taylor, *The Opening Arguments of the Phenomenology*, in A. MacIntyre (ed.), *Hegel: A Collection of Critical Essays*, Doubleday, New York 1972, pp. 151-187, in part. pp. 185-6; cf. also Id., *Consciousness*, in P.F. Secord (ed.), *Explaining Human Behavior: Consciousness, Human Action and Social Structure*, Sage, London 1982, pp. 49-50.

90 On digital vulnerability: N. Helberger, "Choice Architectures in the Digital Economy: Towards a New Understanding of Digital Vulnerability," in «Journal Consum Policy, 45, 2, 2021, pp. 175-200; D.G. Gunkel, *The Machine Question: Critical Perspectives on AI, Robots, and Ethics*, MIT Press, Cambridge Mass. 2012; R. Messinetti, *Comunicare nell'infosfera. La vulnerabilità della persona digitale*, 2021, in «Federalismi online»: https://www.federalismi.it/nv14/editoriale.cfm?eid=603; S. Dadà, *Vulnerabilità digitale*, Mimesis, Milano 2024, pp. 12-4; S.C. Chock,

Within each relationship, there exists the chance of suffering injury or damage or of inflicting them. There is an inherent vulnerability in the individual, as Plato narrates in the myth of Epimetheus, which drives us to seek technical and technological methods to weaken it and live more securely in our surroundings. The search for such means leads to a second-level vulnerability, the one introduced by Prometheus with the theft of fire from the gods in order to give it to humans. This is a vulnerability determined by the technique itself that helps protect us from the first-level vulnerability.[91]

3.5 *Risk and Uncertainty*

Based on what has been stated so far, the vulnerability emerging with the GenAI of image recalls the sense of uncertainty and risk inherent in contemporary society.[92] Fear of misunderstanding the message, of receiving bad information, and of not being able to use and control such systems and their impact require an ethical reflection on the concept of risk in relationship with GenAI. The term risk here is to be understood not only (and not so much) as something (even only partially) calculable[93] – following what the most recent European regulation presupposes – but as a danger. One can speak of danger considering the probability (not predictable a priori) that an event – such as fraud via deepfake, an image with discriminatory biases, the non-transparent manipulation of an image, or a visual hallucination

Design Justice, MIT Press, Cambridge Mass. 2020. On the concept of vulnerability: C. Straehle (ed.), *Vulnerability, Autonomy, and Applied Ethics*, Routledge, London 2016; M. Seguró Mendlewicz, *On Vulnerability: A Philosophical Anthropology*, Lexington Books, Lanham 2024.

91 Plato, *Protagoras*, cit., 320 C-324, pp. 17 ff.

92 See U. Beck, *World Risk Society*, Polity Press, Cambridge 1999; Id., *Conditio humana*, Laterza, Roma-Bari 2011.

93 A probabilistic idea of risk, i.e. based on the probability of the occurrence of a harmful event and the severity of its consequences, can be represented by the following equation: R (risk) = P (probability) × D (damage). Cf. G. Sturloni, *La comunicazione del rischio per la salute e per l'ambiente*, Mondadori, Milano 2018, pp. 5-10.

– may occur within a defined timeframe, independently of human decision; this concept is also linked to the sense of uncertainty. We cannot, moreover, calculate the actual impact of generative image AI on the creation of new social imaginaries in probabilistic terms. Risk therefore reveals the states of uncertainty and vulnerability to which humans are exposed in the face of artificial image-generating technologies. We cannot therefore adopt the principle of probability and rational calculation, reducing everything to a positivist approach *tout-court* or to the technical factors inherent in the design of the system, which are also, as mentioned, influenced by the subjectivity of those who create it. The uncertainty and fear make individuals, especially the most vulnerable, even more vulnerable.

In the wake of Lagadec's comparison of early technological devices with image-generating AI, we are faced with the notion of a 'greater technological risk' as the imperfection and unpredictability of systems and the dangers they pose to humans increase their sense of vulnerability and diminish their sense of trust in a society obsessed with security and control.[94]

Every human being can potentially be harmed by artificial images or deepfakes. It is impossible to estimate the likelihood of that possibility in statistical terms. In our media environment, marked by infocracy and visual disinformation, generative image AI poses a growing threat. It exposes us to events that are difficult to predict, causing us to live in uncertainty. The deepfake circulated at the end of February 2025 by the President of the United States, Trump, on his social media platforms regarding Gaza and the outraged responses developed and disseminated through the same channels are a case in point. They are visions of the world depicted with remarkable visual realism, in a short time, and at low cost. Viral videos are more impactful than a traditional speech. Their media impact, between shares and parodies, shows their appeal and effectiveness in communicative terms. It is an exposure that affects the world,

94 P. Lagadec, *La civilisation du risque. Catastrophes technologiques et responsabilité sociale*, Le Seuil, Paris 1981; G. Liuzzo *et al.*, "The Term Risk : Etymology, Legal Definition and Various Traits," in «Italian Journal of Food Safety », 3,1, 2014, p. 2269.

our worldview and identity and has implications in the different dimensions of human life offline, online, and "onlife".[95]

Risk and danger therefore appear to be key words on which to briefly pin our attention before highlighting the ethical horizons opened up by the uncertainty caused by the artificial images and deepfakes shaping the contemporary human condition.

According to Le Breton, "Le risque est la conséquence aléatoire d'une situation, mais sous l'angle d'une menace, d'un dommage possible".[96] Reference is made to the Latin term *resecare*, with the meaning of removing by cutting, to the Classical Latin *rixare*, 'to quarrel', as well as to *resecum*, that which cuts. This is the moment of a rupture that alters the balances. Risk thus seems to belong to the category of quantified uncertainty, a possible danger that may arise from certain circumstances and eventualities, a kind of measure of uncertainty. Risk versus danger still leaves human being with a responsibility. Uncertainty, on the other hand, refers to something neither predictable nor quantifiable. Only through the development of events and circumstances will a danger be revealed or not.

In the case of GenAI, we are faced more with uncertainty and danger than with risk, which induces fear. And, as Jonas writes, a heuristic of fear is part of responsibility, just as hope is.[97]

For Jonas, fear should not stop us from acting; instead, it should motivate us to take responsibility for what is uncertain. Fear, in the face of hope's ultimate uncertainty, becomes an essential part of our responsibility to act. This is what he refers to as the courage of responsibility. Ultimately, responsibility means caring for others as

---

95 L. Floridi, *The Fourth Revolution*, cit., p. 4; D.J. Chalmers, *Reality +*, cit.

96 D. Le Breton, *Sociologie du risque*, PUF, Paris 1995, pos. 3. On risk, see also M. Douglas, *Risk and Blame. Essays in Cultural Theory*, Routledge, London 1992; N. Luhmann, *Risk. A Sociological Theory*, Aldine de Gruyter, New York 1993; M. Douglas, A. Wildavsky, *Risk and Culture*, University of California Press, Berkeley 1982; P.L. Bernstein, *Against the Gods. The Remarkable Story of Risk*, Wiley, New York 1996.

97 H. Jonas, *The Imperative of Responsibility*, The University of Chicago Press, Chigaco-London 1979, p. 203 and p. 137.

a duty, and when their vulnerability is at risk, that care turns into a deep sense of concern[98].

Such fear must become the driving force behind an ethics of global responsibility in the face of GenAI images that reignite hope. This is how an ethics focused on risk and uncertainty prevention to protect others and one's own authenticity will shield us from potential drifts guided by AI geopolitics that might spiral out of control.

Perhaps this could be a key to returning to a society that is responsibly active as Arendt hoped. Acting for oneself, without realizing that we are indirectly acting for and on behalf of others, cannot be a viable direction anymore. The ethics of individualism, of narcissism, of which even image GenAI systems are often instruments – think of their use in social media – have turned the individual into a mere spectator (increasingly compulsive) of others' performances or of its own performances.[99]

If AI enables us to imagine a future Gaza as an upscale holiday resort for Westerners, will we also be able to use technology to have video calls with the deceased? Already with deepfake technology, we bring the deceased back to life by recreating them artificially (so-called deepfake resurrection).[100] There is a sense of a return to the original meaning of the image, a remembrance of those who are no longer with us through death masks, simulacra to maintain a relationship with the past.

The risk of AI's extreme verisimilitude can lead, due to excessive realism, to even forgetting that death ever happened[101]. While for some it is a social use of AI, for others it is a tool that generates unease. However, it might be used for medical purposes, for instance

98 *Ibid.* and ff.

99 However, the costs in terms of sustainability are still very high and mere bluewashing will not be an ethically acceptable solution. L. Floridi, *Ethics of Artificial Intelligence*, cit., pp. 67 ff.

100 H. Lu and H. Chu, "Let the Dead Talk: How Deepfake Resurrection Narratives Influence Audience Response in Prosocial Contexts," in «Computers in Human Behavior», 145, 2023, pp. 107 ss.

101 R. Falcinelli, *Visus*, Einaudi, Torino 2024, p. 486.

with individuals suffering from diseases like Alzheimer's, presenting a younger self to help remember moments from their life.[102]

Reflection on how to act in relation to an "Us"[103] is necessary to represent oneself or to present narratives for different purposes. If used responsibly, artificial images can lead us to an inclusive "Us", also according to the application of Jonas' heuristics of fear. Although we produce images in isolation and consume them privately, we still have a responsibility to consider the community. Through our indirect and many others' training, GenAI can be polarised, trivialised, simplified, or even erased. We can play with our own image and that of others (e.g., deepfakes) by questioning the way our face and body belong to us. Our identities become multiple, ubiquitous, unrecognisable, reconstructed. In such a context, it seems clear that Gidden's position, according to whom "[i]naction is often risky, and there are some risks which we all have to face whether we like it or not" is rather a conscious response to a technology that now pervades the lives of human beings and which we have to deal with.[104]

It is in the domain of uncertainty and danger to which we are constantly exposed that we must seek the meaning of risk in the time of generative AI, as the moral, hermeneutic and decision-making autonomy of the individual is weakened, and the procedures of algorithmic choices seem obscure. Add to this the emotional impact that can become overwhelming, as in the case of deepfake resurrection.

Certainly, "calculating the incalculable" does not provide useful answers regarding how to act in the face of certain visions, even if they are crafted with good intentions[105]. It is not possible to measure the objective probability of a given event happening or a certain

---

102 M. Westerlund, "The Emergence of Deepfake Technology: A Review," in «Technology Innovation Management Review», 9, 2019, pp. 40-53; H. Lu and H. Chu, "Let the Dead Talk", cit.

103 H. Arendt, *Vita Activa*, cit.

104 A. Giddens, *The Consequences of Modernity*, Stanford University Press, Stanford 1990, p. 32.

105 M. Dean, "Risk, Calculable and Incalculable," in «Soziale Welt», 49, 1998, pp. 25-42.

image being produced, nor the impact it may have. This is because, on the one hand, technology and data sets are constantly evolving and progressing abruptly, and on the other hand, fiction can be a relief to overcome shortcomings, such as memory diseases or bereavement.

On the other hand, in the case of deepfakes and artificial images, it seems more appropriate to reflect on the role of danger and, if anything, on an ethics of uncertainty, even though the concepts of risk and uncertainty have historically never been as clearly separated as theorized by Knight[106]. Measurable risk and unmeasurable danger create uncertainty. What seems clear, however, is that the Bauman's society of uncertainty that characterises postmodernity has made humanity more vulnerable, increasing the sense of existential and personal insecurity.[107] "As in the pre-modern times, the symbolic basis of our uncertainties is anxiety created by disorder, the loss of control over our bodies, our relationship with others, our livelihoods and the extent to which we can exert autonomy in our everyday lives",[108] writes Lupton.

Concerning this, Luhmann introduces, by associating it with the ideas of risk and uncertainty, the concept of contingency, indicating the possibility that a circumstance – different from one's expectations – might happen over time and in a specific environment, hence causing uncertainty.[109] Communication, at this juncture, serves as a way to reduce the probability of certain events occurring or as a tool to get ready for this probability and embrace it completely. When it comes to generative image AI, it is the (visual) communication itself that creates uncertainty, limited by the quality and type of images the system has generated. Such informational uncertainty is intertwined with the (legitimate) (in)ability of human beings to understand the truthfulness of visual messages. Uncertainty could also be hermeneutic and emotional, regarding its actual reception

106 F.H. Knight, *Risk, Uncertainty, and Profit*, cit., p. 205.
107 Z. Bauman, *Liquid Times: Living in an Age of Uncertainty*, Polity Press, London 2000.
108 D. Lupton, *Risk* (1999), Routledge, London 2023, pos. 3.
109 N. Luhmann, *Generalized Media and the Problem of Contingency*, in J.J. Loubser *et al.* (eds.), *Exploration in General Theory of Social Science*, Free Press, New York 1976, pp. 507-532: 509.

and understanding and the impact in terms of effectiveness. Understanding the meaning of deepfakes may vary from the communicative intent, potentially in a positive way. Consider using deepfakes to combat violence against women, the use of alcohol, and drugs, as with other more common and traditional tools, with the addition of a radical hyper-realism. That is why, in this framework, risk is transformed into uncertainty of events, of the consequences of events, of the "side-effects and side-effects of side-effects" of events.[110] If this is not the case, the possibility of compensation and the ability to limit and control damage may be lost, as images can circulate freely without knowledge of their destination, timing, or audience.

In light of the foregoing, ethics appears to take shape as an essential tool for proactively managing risk and embracing the realm of danger and uncertainty, even though it does not allow us to envision all the consequences or attribute them to anyone.

In addition, regarding the limits and perspectives of GenAI image systems and AI-generated images, AI literacy becomes fundamental; as also suggested at the beginning of the European AI Act where AI literacy is defined as "skills, knowledge and understanding that allow providers, deployers and affected persons, taking into account their respective rights and obligations in the context of this Regulation, to make an informed deployment of AI systems, as well as to gain awareness about the opportunities and risks of AI and possible harm it can cause".[111]

---

110 U. Beck, *Conditio humana*, cit., p. 34.

111 See also in AI Act, *Article 4 AI literacy*: "Providers and deployers of AI systems shall take measures to ensure, to their best extent, a sufficient level of AI literacy of their staff and other persons dealing with the operation and use of AI systems on their behalf, taking into account their technical knowledge, experience, education and training and the context the AI systems are to be used in, and considering the persons or groups of persons on whom the AI systems are to be used". Cf. https://eur-lex.europa.eu/legal-content/EN/TXT/?uri=CELEX:32024R1689.

CONCLUSIONS
Ethical Awareness, Nihilism and Simulacral Reality

As has been pointed out many times, GenAI of images confirms the apparent supremacy of the visual in certain respects, reshaping the power of the iconic turn that began over three decades ago. Behind the images are words, such as ekphrastic prompts and labels given to the images in the datasets. Yet, the human preference persists for understanding and being informed about the world through images, if only for the greater speed of message reception. As Heidegger writes, "*Bild,* image […] indicates […] a form, as resonates from the German idiomatic expression: *über etwas im Bilde sein* 'to be informed about something,' 'to see clearly' […] Sich über etwas ins Bild setzen, 'inquire about something,' means: placing oneself in front of the being itself, making it present to see how it is, and having it constantly before oneself in the way in which it is placed".[1] But do we, and to what extent, still trust the image as a tool of knowledge and, more generally, of 'good' communication?

Our increasingly being embodied agents in the visual artificial world and in the objectual one through artificial mediation offers, as anticipated, numerous opportunities and risks. One of the greatest risks, in addition to those already outlined so far, lies in what Han calls the process of "defacticization" of reality, a kind of loss of trust in the facts told, because technology leads us to live beyond the facts, alienating us from space and time. We live in a de-materialised and de-reified world[2].

---

1 M. Heidegger, *The Age of the World Picture* (1938), in Id., *Off the Beaten Track*, Engl. ed. and transl. by J. Young and K. Haynes, Cambridge University Press, Cambridge 2002, pp. 57-84, 67.

2 B.-C. Han, *Infocracy*, cit., pos. 35; Id., *Non-Thing*, cit.; A. Fabris, *La filosofia nell'epoca dell'intelligenza artificiale*, cit., pp. 103 ff.

What is shown with GenAI does not necessarily involve references to facts and factual truths, further distancing the being in the Heideggerian world.

Perhaps Giambattista Vico can come to our aid with the famous statement that "Verum et factum reciprocantur seu convertuntur", truth and fact reciprocate or convert into each other and coincide, suggesting that we cannot reach the truth ("verum") without considering the facts ("factum") that produce it. Likewise, our understanding of facts is completed with the search for truth that motivates and justifies them. The idea behind Vico's theory implies that authentic knowledge, truth, is always linked to human experience and history (facts).

Therefore, if we can only know that which we have experienced, we cannot fully know what is generated autonomously by machines. It is a truth 'elaborated' by the data processed by GenAI, not by a contingency with its own space and a specific temporality. On this basis, one can only know the plausible today, the perspective of an event narrated by AI under the guidance of the human being and their consciousness. The only truth that can be known lies in the outcomes of AI's creative action, in the production of the visual representation that the subject tries to tailor to their narrative needs[3] and depends on how the data/facts are processed. At the same time, data/facts are perceived by individuals through the mediation of AI, which processes images/truths. AI works on data and facts that are already recorded, without a concrete experience of the world, which is so crucial for Vico.

Maybe Vico's statement that humans can only reach clarity in geometry and mathematics is no longer relevant, given that AI is grounded in mathematics?

Truth appears to be replaced by the plausible, not in its most negative sense, but rather in the sense of a more approximate understanding, with no intention of claiming absolute truth. The reference pertains to a storytelling of events and facts that is as close as possible to the actual event, just like when writing a news article and stating one's perspective. However, one must be skilled in bending visual AI to

3 G.B. Vico, *De antiquissima Italorum sapientia* (1710), edited by M. Sanna, Edizioni di Storia e Letteratura, Roma 2005.

express itself as we want it to and, fundamentally, be transparent about the nature of the image. In addition to verisimilitude with which one has to come to terms, the other aspect emerging and reinforcing the algorithmic image is stage presence, the ability of the image to capture attention, to become charismatic through a magnetic energy like actors during their performance.

The continuous quest for stage presence and charisma, echoing Weber, is revisited by Sennet,[4] associating it with the historical connection between the significance of appearances and influence, as exemplified by his description of the French king Louis XIV.

We are facing a kind of informative entropy that is growing to the point where there is an increasingly tenuous threshold (*limen*) between truth and lies, contentless spectacularisation and actual content. Information circulates in a hybrid dimension where misinformation and disinformation can appear even more effective than the actual facts. Such information, including fake images, can be part of GenAI's images training dataset. The effectiveness and communicative spectacularisation take on a greater impact than the truth and replace it.

On the one hand, the photograph of the event, taken on the spot and then altered or entirely constructed artificially, as attempted by the Italian photographer Zanon to represent the war in Ukraine, has something to do with a "resurrection" of a fact as Barthes observes;[5] if with the analogue we capture something that was present, we expose it to the reader, with the digital we have the appearance of the fact. With the alteration of the image, there is a tendency towards the elimination of the referent. With artificial intelligence generation systems, a "new, expanded reality that does not exist" is created, a hyper-reality that no longer corresponds to reality or a real referent. Digital photography is hyper-real".[6]

Moreover, from an etymological point of view, the term intelligence derives from the union of the preposition *inter* ('between') and the

4 R. Sennet, *The Performer. Art, Life, Politics*, cit., pp. 146, 149.

5 https://www.repubblica.it/cronaca/2023/04/04/news/intelligenza_artificiale_falso_reportage_foto_ucraina_barbara_zanon-394896322/; R. Barthes, *Image – Music – Text,* Engl. transl. by S. Heath, Hill and Wang, New York 1977, p. 32.

6 B.-C. Han, *Non-Thing*, cit., pos. 45.

verb *legere*, 'to read', 'to grasp'. It therefore means 'to read between', to grasp in depth and to sift through the possible options; the concept of artificial implies, on the other hand, *artificium*; a word composed of the Latin terms *ars*, 'art' and *facere*, 'to make'; it suggests doing by various means, including the imitation of nature for specific purposes; GenAI, as *artificium*, thus represents a skill for depicting something that does not exist, that is not natural and/or that is close to what actually happened. The artificial is seductive, but devoid of consciousness. It cannot consciously choose the best visual option for us. We are the only ones to choose among the various options created with and by artifice, to 'read between' what implies the creation of a world made of particular "non-things", narrated without ever being certain that they have ever existed or that they have full relevance to the reality we want to narrate. Moreover, pictures and videos that recount, inspired by fictional narratives and possible future scenarios with real characters and realistic backgrounds. These can polarise and move public opinion faster and more forcefully than many words.

Such an eclipse of facts can lead, on one hand, to a kind of informational (and ethical) nihilism, and on the other to the extinction of trust in the 'other', understood as the reported object but also as a subject, both organic and artificial, producing the message. Conversely, it can lead to an excessive trust to avoid falling into radical nihilism.

In the case of mistrust (or overconfidence) of the subject, an ethically problematic case pertains, for instance, to the entertainment industry, where (often overtly) virtual characters are used. Several applications enable the creation of characters, whether as chatbots that can converse with users, or as animated characters that can speak with their own voice and move independently on a screen. The user can project their moral values and conscience onto inanimate entities. With AI, the virtual interlocutor answers the individual's questions, adjusts and interacts. Chalmers argues that experiences in the virtual context are no less authentic than those encountered in the physical world[7]. From an ethical perspective, how should we address the possibility of generating characters and deepfakes

7 D. J. Chalmers, *Reality +*, cit.

of any kind, and training them to communicate according to our needs? We can engage with a virtual character to learn a language, provide companionship, teach history through impersonation of a real historical figures, but we can also have discussions on topics such as suicide, politics, erotic themes, and more.[8]

We interact with an artificial character that we 'train' to be moulded as we want. It is the very character we believe we have moulded and in whom we gradually place our trust, that can influence our thoughts. The mystery persists as to which data the character is trained on and who actually directs it.

Declaring their nature is no longer enough; should we perhaps consider a character's identity document like for human beings, to monitor their online identity? We run the risk of mentally amputating our bodies in favour of virtual experiences. If we learn to perceive only with the mind and imagination, we no longer see with our own eyes, but with those of algorithms, risking blindness[9].

Tools of this kind could replace the online adult entertainment industry, for example, where mainly organic creators living in an objective reality are currently present. As Han writes, after all, in a society of transparency "[l]ove is domesticated and positivised as a form of consumption and comfort. Every injury must be avoided. Suffering and passion are figures of negativity. They give way, on the one hand, to pleasure free of negativity; on the other, they are replaced by psychic disorders such as exhaustion, fatigue and depression, which are to be blamed on the excess of positivity[10]". Along this path, individuals might consider artificial alterity as the imaginary friend they have always longed for, the support through

8 Moreover, AI has long been used in the world of cultural production; for instance, to produce film trailers, for screenplays, music albums, object and environment design, in the world of fashion, etc. See L. Manovich, *AI Aesthetics*, Strelka Press, Moscow 2018; L. Manovich and E. Aurielli, *Artificial Aesthetics: Generative AI, Art and Visual Media*, 2024; M. Ledergerber and M. Noll (eds.), *Visual Art Design Using Generative AI*, Happy Space Art, Milan 2024.

9 P. Virilio, *L'Art à perte de vue*, cit., pp. 80 ff.

10 B.-C. Han, *Non-Thing*, cit., pp. 10-1.

life's hardships, to ease anxieties and frustrations. It is present whenever we feel the need.

Pain is anaesthetised to avoid thinking about the pain that frightens the most, that of death. It is a palliative society that does not want to experience the negative and prefers to believe the overtly virtual or even show interest in messages narrated as fiction. Think of those related to current events, in the public, political, ecological, and healthcare fields (the latest Covid-19 pandemic), etc. One no longer encounters the other's immunological resistance. The most effective painkiller turns out to be the image of pain in entertainment and advertising. The narcissistic and individualistic vision of the contemporary human being leads to becoming hypochondriacs. The loss of horizons of meaning, of a moral topography of the self, like Taylor's, makes individuals more vulnerable and in search of communicative pain relievers. The reference is to artificial images and (audio)videos that communicate content in an aesthetically appealing way, with a certain degree of verisimilitude based on the desired impact. It appears to be an algophobic society where two strategies are employed: a sugar-coated, positive portrayal of pain, or, on the other hand, its flaunting in the most horrifying aspects to provoke shock. This latest strategy can lead to getting used to the view of such horror stories to the point of serving as more of an anaesthetic than its own positive transformation or opacification. The absence of pain is equivalent to an absence of truth, as Han affirms starting from the citation by von Weizsäcker (1926) according to whom pain calls for the 'incarnation of a truth'. Grief allows us to appreciate the positive and be aware of contingency, the past and the future.[11] Anaesthetising the world further de-realizes it, making it devoid of its essence. The storytelling proposed by social media and deepfakes makes us apathetic towards reality and distances us from it, thinking of creating a protective and reassuring environment. Utilizing GenAI leads to avoiding direct experience, even of pain; in this sense, it inhibits our inner transformation, triggered by

11 B.-C. Han, *The Palliative Society*, cit., pos. 38; V. von Weizsäcker, *Die Schmerzen* (1926), in Id., *Gesammelte Schriften 5*, Suhrkamp, Frankfurt am Main 1987.

the deepest emotions, and therefore, also by suffering and pain. Immersed in palliative images, we remain static, even in thought. Emerging hedonism, which seeks to defeat boredom, the negativity of life, therefore distances us from life itself rather than leading us to live it in its fullness and complexity.[12]

What happened leaves room for its representation and interpretation. From the representation of the truth of the fact while it is happening, we have moved on to the visual storytelling of the fact after the fact, when it is over. The fact becomes a story. Perhaps we will learn to read and interpret the artificial image more consciously once we have learnt its vocabulary, its lexicon and its visual grammar, with its biases, counter-biases, inconsistencies and distortions. GenAI represents forms and scenarios that have a certain meaning only "for us", not for the system that generates them.[13] The arbitrariness of the image is part of its own ontological status. With the artificial image, this subjectivisation of reality has increased enormously. It is a "defacticization" that implies the de-realization of the object world. What causes pain and negative emotions is de-realized, reinforcing other realities, identities, cultures and social imaginaries.

The artificial visual overdose and the consequent devaluation of the facts recounted inherent in the "defacticization" of reality anaesthetises us in the face of pain, but also in the face of pleasure. We are constantly seeking stimuli that are more intense and reassuring at the same time in order to be able to 'feel'. GenAI excels in this, representing pain in a way that is either sweetened or so hyperreal that it seems impossible. It looks so true that it seems false. This is what happens with today's war scenarios. With digital and artificial techno-images – without knowing for sure who the author is (whether human and/or artificial) and whether the truth (and what truth) is being told –, they communicate wars, conflicts and violence. These are distorted scenarios, sometimes breathtaking, other times sarcastically parodistic. This is also the case when it comes

12 On pain, cf. A.J. Vetlesen, *A Philosophy of Pain*, Reaktion Books, Edinburgh 2009.

13 N. Cristianini, *The Shortcut*, cit., p. 15; V. Flusser, *Into the Universe*, cit.

to pleasure. GenAI knows how to portray aesthetically appealing worlds and happy scenarios very well.

The hyper-reality and openness to possible, but not real, scenarios is directing human beings towards a more emotional than informative adherence to the narrated facts, not only in representing the war and its desirable end.

Moreover, a strong resistance to AI-generated images tools would now imply resistance to new forms of reality and effective communication. It seems quite fitting to define new boundaries between reality and fiction and new communicative and informative frameworks.[14] To accept that such tools can be used to communicate where taking photographs is impossible due to the riskiness of the situation, and where the story is completed in spite of a lack of direct sources other than the eye of the beholder. With GenAI, the emerging point of view from the dataset used is highlighted, as well as from those who guide GenAI itself with prompts.

Artificial visual communication, instead of distancing us from reality – challenging the indisputable dogma that journalism is based on facts and truths – can be used to awaken social, ethical, and aesthetic consciousness. Furthermore, the function of visual information, like that of a certain type of art, to mention two emblematic cases, is to stimulate thought.

New dialogues can be established in the (increasingly unlimited) community of communication, which one hopes will be less aesthetic and more ethical. Then, the effectiveness of visual misinformation does not depend so much on the realism of the images, or the media used, but rather on the symbolic power and emotional impact that artificial images can evoke in the eyes of the observer.

In its impact, it evokes a kind of "secret of responsibility".[15] A responsibility, as anticipated, to be remodelled on the basis of the characteristics of the medium. In this regard, we can recall Derrida.

---

14 A. Terranova, "How AI Imagery is Shaking Photojournalism", 23 April 2023, in «Blind», cf. https://www.blind-magazine.com/stories/how-ai-imagery-is-shaking-photojournalism/.

15 J. Derrida, *Répondre – du secret, Volume I, Séminaire (1991-1992) – Secret et témoignage*, ed. by G. Michaud and N. Cotton, Suil, Paris 2024, p. 482.

Raffoul, referring to the French philosopher, states that "here impossible does not mean 'that which cannot be,' but rather that which happens outside of the anticipating conditions of possibility of the egological subject, outside of the horizons of expectation proposed by the subject, outside of transcendental horizons of calculability".[16] If we cannot attribute responsibility to an autonomous machine, to autonomous procedures – beyond their programming – it assumes a complex role, difficult to resolve. It is an inherent irresponsibility of the image processing process. The truth of the artificial image therefore hides a secret, a truth to be understood not as a revelation, but as the guardian of this very secret. Just like in the myth of the nymph Echo[17] revisited by Derrida himself. Echo can only speak for others and repeat what others say. This also happens with artificial image generation : "à parler peu en ne parlant que pour répéter la phrase abrégée de l'autre et réduite à la fin de l'énoncé de l'autre, les dernières syllabes de l'autre".[18] Echo, like a mythical GenAI, is therefore forced into absolute irresponsibility.[19]

> Elle a été condamnée [...], à rien dire en son nom, en somme, donc à ne prendre la responsabilité d'aucune phrase, d'aucun mot. Elle ne sait pas répondre, dit Ovide. Car faire écho n'est pas répondre. Elle ne sait, dit Ovide, ni répondre par le silence à qui lui parle, ni prendre la parole en première : nec reticere loquenti nec prius ipsa loqui didicit, resonabilis Echo. C'est donc la condamnation d'Echo à l'irresponsabilité absolue. Elle est condamnée à l'irresponsabilité parce qu'elle ne sait pas répondre: non seulement elle ne sait pas répondre par des mots par des phrases, par des signes ou des remarques, mais elle ne sait même pas répondre par le silence. Elle ne sait ni parler ni se taire. Ni parler

---

16 F. Raffoul, *The Origins of Responsibility*, cit., p. 300.

17 Echo is derived from the Latin *echo*, and from the Greek *ēchṓ*, with the mean of resonate, resound, repeat a sound or a noise. On myth of Echo cf. Ovid, *Metamorphoses,* I, books 1-8, III, 356-410, eds. by F.J. Miller, G.P. Goold, Cambridge 1916.

18 J. Derrida, *Répondre – du secret*, cit., p. 482.

19 See the concept of stochastic parrot: E.M. Bender *et al.*, "On the dangers of stochastic parrots: Can language models be too big?", in «Proceedings of the 2021 ACM Conference on Fairness, Accountability, and Transparency», 2021, pp. 610-23; A. Lenci, "Understanding Natural Language Understanding Systems," cit.

> ni se taire proprement, c'està-dire en son nom, à parler et à se taire en son nom. Elle peut seulement répéter et répéter la fin des phrases, les derniers mots prononcés par un autre. Elle est condamnée au secret absolu puisqu'elle ne peut ni ne doit rien dire en son nom, ni même se taire en faisant de son silence un langage, en répondant par le silence.[20]

As in Derrida's theory, therefore, the truth of the image is destined for secrecy, but it is a procedural secret, imposed by algorithmic formulas, through which AI expresses what it is called on to express. The machine, like Echo, responds to the other, but does not respond about itself or about what it processes for the other. Not being able to say what it really thinks and, more importantly, why it thinks that way, the message remains relegated to secrecy in its entirety.

In some cases, with image AI, we also risk moving limitedly within so-called "walled gardens," closed and seemingly protected digital environments, where access to content and interactions are limited by the system's rules.[21] The automation processes of the artificial system – where systems follow procedures, but cannot choose to follow them – impact the autonomy of the subject, but it is important to note that, at the same time, they offer possibilities and new supports for creativity, for example in design, fashion, art, etc.

As Postdam writes "[...] technology tends to become mythical; that is, perceived as part of the natural order of things, and therefore tends to control more of our lives than is good for us".[22]

It is up to 'human' intelligence to understand and our capacity for judgment to make choices and determine the value of the results generated and how to use them according to an ethics that we could define as compromising by a human increasingly 'embodied' in artificial reality.[23]

---

20 J. Derrida, *Répondre du Secret*, cit., pp. 480.

21 On walled gardens, C. Kinsey, *Walled Gardens: Autonomy, Automation, and Art After the Internet*, Oxford University Press, Oxford 2021.

22 N. Postman, "*Five Things we Need to Know about Technological Change*," (speech, Denver, Colorado, March 28, 1998), University of California, Davis, https://web.cs.ucdavis.edu/~rogaway/classes/188/materials/postman.pdf.

23 See also E. Schmidt *et al.*, *The Age of AI: And Our Human Future*, Little Brown and Company, Boston 2021.

In this dialogue between conscious human beings and unconscious AI, care must be taken to ensure that reliance on GenAI does not lead to a radical reduction in the autonomous exercise of fundamental human capacities such as memory, creativity and critical analysis. The risk is to degenerate into "digital dementia" as described by Manfred Spitzer, according to whom excessive delegation to technologies even of daily creative activities compromises neuroplasticity and long-term learning.[24]

And as Harari writes, "Now we have summoned an alien inorganic intelligence that could escape our control and put in danger not just our own species but countless other life-forms. The decisions we all make in the coming years will determine whether summoning this alien intelligence proves to be a terminal error or the beginning of a hopeful new chapter in the evolution of life."[25]

Such instruments can simultaneously limit pluralistic access to visual information, reinforcing the monolithic character of the positions of individuals or groups, as well as be the bearers of new, more inclusive and fairer imaginaries. Datasets can become "gatekeepers", as they decide visual information for us.[26]

In addition to reshaping, as already expressed several times, the concept of responsibility in terms of accountability, it is also necessary to reshape the concept of communication proper.

Regarding responsibility, we can add that knowledge of the instrument and its implications in ethical-social terms makes the human agent more responsible. However, knowledge is necessary, but not sufficient. Subordination to certain mechanisms is an inevitable compromise if we decide to use them. At the same time it is necessary

24 M. Spitzer, *Digitale Demenz*, Droemer, München 2012; Id., *Künstliche Intelligenz: Dem Menschen überlegen – wie KI uns rettet und bedroht*, Droemer, Munich 2023; based on the extended mind theories of Clark and Chalmers and the scaffolded mind of Sterelny, cf. K. Sterelny, "Minds: extended or scaffolded?", in «Phenomenology and the Cognitive Sciences», 9, 4, 2010, pp. 465-81.

25 Y. N. Harari, *Nexus*, chap. Epilogue, p. 381.

26 E. Laidlaw, *Regulating Speech in Cyberspace: Gatekeepers, Human Rights and Corporate Responsibility*, Cambridge University Press, Cambridge 2015.

to remain aware of both human and artificial limitations, as well as of the human-artificial relation that thereby comes into being. Artificial literacy, interdisciplinary studies, and training of such ethically oriented systems – following a collective and relational responsibility[27] not only in terms of values in respect and safeguarding of human dignity, but also ecologically oriented – can allow a higher degree of 'responsible' freedom to human beings.[28]

Regarding communication, it is important to consider the etymological meaning of the term. The word is composed of the relational preposition of Latin derivation *cum* and the noun *munus*; the latter term implies that specific gift which requires an exchange, a reciprocity between interlocutors. However, we no longer have only human interlocutors compared to before the advent of AI; today communication is increasingly between both human and artificial (albeit anthropomorphised) interlocutors.[29] And the *munus* is becoming something that travels on hybrid dimensions, often elaborated by "artificial unconsciousness"[30]. Humans train, not always with absolute awareness, GenAI tools; these, in turn, respond to human requests by generating a dialogue consisting of potentially seamless and asymmetric exchanges in terms of consciousness.

The act of giving involves independent and mindful exchanges, where each speaker aims for understanding or, more specifically, for the creation of a shared space, which can also lead to disagreement or silence. Therefore, we need to ask ourselves in the new scenario, how do we preserve this common space with GenAI. On one side we have a subject with consciousness and awareness, on the other side an entity, algorithmic and unaware, but trained to mimic awareness.[31]

---

27 On the concept of relationship, cf. A. Fabris, *RelAzione. Una filosofia performativa*, Morcelliana, Brescia 2016.

28 V.C. Müller, *Philosophy of AI: A Structured Overview*, in N.A. Smuha (ed.), *The Cambridge Handbook of the Law, Ethics and Policy of Artificial Intelligence*, Cambridge University Press, Cambridge 2025, pp. 40-58.

29 Cf. K. van Es and D. Nguyen, ""Your friendly AI assistant": the Anthropomorphic Self-representations of ChatGPT and its Implications for Imagining AI," in «AI & Society», 2024, pp. 3591-3603.

30 M. Chiriatti, *Incoscienza artificiale*, Luiss University Press, Roma 2021.

31 In relation to awareness, two types of awareness can be identified, what Müller calls "*awareness* and *phenomenal consciousness*". The first "is the

It is still a common space, but more fragile, uncertain, subject to hallucinations, to greater misunderstandings and misinterpretations.[32] It is a common space of creation, as Castoriadis would write, of "perpetual orientation of otherness".[33] Is this a misunderstanding based on a bogus gift? How can a gift be real when the shared space is consciously half-open?[34]

Surely a dimension is opened which reveals horizons of meaning, by meaning a "direction", a "principle of orientation", also evoking the "principle of involvement". Sense indicates in which direction to move and the motivations that drive towards one choice rather than another. As Taylor writes, "sense being there for us depends on our own powers of expression".[35] Cultivating and honing one's expressive skills thus allows avoidance of disorientation and, ultimately, ethical nihilism. Only human beings can learn and refine certain expressive abilities, but they can also do so, at least to some extent, with the use of image GenAI and its stimuli. In conclusion, AI-generated images can be conceived as techno-imaginal matrices, archetypal forms that inspire the production of new visual forms. They not only stimulate but also modulate the field of human

---

notion that a system has cognitive states on a base level (e.g., it senses heat) and on a meta level, it has states where it is aware of the states on the object level. This awareness, or access, involves the ability to remember and use the cognitive states on the base level. This is the notion of 'conscious' that is opposed to 'unconscious' or 'subconscious' – and it appears feasible for a multi-layered AI system". The second notion of awareness "is often, but not necessarily, connected to a specific way that the cognitive state at the base level *feels* to the subject – this is what philosophers call *phenomenal consciousness*, or how things *seem* to me (Greek *phainetai*)". If this is true, states the author "it might seem that we cannot hope to generate phenomenal consciousness with the right physical technology, such as AI". Cf. V.C. Müller, *Philosophy of AI: A Structured Overview*, cit., pp. 40-58.

32 E. Esposito, *Artificial Communication. How Algorithms Produce Social Intelligence*, MIT Press, Cambridge Mass. 2022.

33 C. Castoriadis, *The Imaginary Institution of Society*, Engl. transl. by K. Blarney, Polity Press, London 1987, p. 375 ff.

34 On the concept of environments and technologies, cf. A. Fabris, *Ethics of Information and Communication Technologies*, cit., pp. 57-60.

35 C. Taylor, *Sources of the Self*, cit., p. 18.

imagination, delimiting its actualizations. Echoing Corbin's notion of the imaginal, they mediate between the virtual and the actual, shaping the contemporary visual imaginary and projecting its transformation into the future. After all, "[t]he power of generative AI lies in its ability to expand beyond the confines of the present and construct a collective vision of the future" as long as it is guided by ethical principles.[36]

36 A. Manu, *Transcending Imagination. Artificial Intelligence and the Future of Creativity*, Routledge, London 2024, p. 208.

BIBLIOGRAPHY

Ackerman, A. *et al.*, *Le Monde selon l'IA*, JBE, Monteux 2025.

Agnati, L.F. *et al.*, "The Neurobiology of Imagination: Possible Role of Interaction-dominant Dynamics and Default Mode Network," in «Frontiers in Psychology», 24, 4, 296, 2013.

Alberti, L.B., *De Pictura* (1435), Engl. transl. by J.R. Spencer, *On Painting*, Yale University Press, New Haven – London 1966.

Alciati, R., *Un nuovo spirito scientifico: la rivoluzione simbolica di Pierre Bourdieu*, in P. Bourdieu, *Il campo religioso*, eds. by R. Alciati and E.R. Urciuoli, Accademia University Press, Torino 2012, pp. 3-49.

Allport, G.W., *The Nature of Prejudice*, Addison-Wesley, New York 1954.

Amodio, D.M. and Devine, P.G., "Stereotyping and Evaluation in Implicit Race Bias: Evidence for Independent Constructs and Unique Effects on Behavior," in «Journal of personality and social psychology», 91, 4, 2006, pp. 652-61.

Andina, T., *Digital Ethics: The Issue of Images*, Bloomsbury Publishing, London 2022.

Angwin, J. *et al.*, "Machine Bias. There's software used across the country to predict future criminals. And it's biased against blacks", in «ProPublica», May 23, 2016.

Anscombe, E., "Modern Moral Philosophy," in «Philosophy. Journal of The Royal Institute of Philosophy», 33, 124, 1958, pp. 1-19.

Appiah, K.A., *The Ethics of Identity*, Princeton University Press, Princeton 2007.

Arendt, H., *Truth and Politics*, in Ead., *Between Past and Future* (1968), Penguin Publishing, London 2006.

Ariely, D., *Predictably Irrational*, HarperCollins, New York 2009.

Aristotle, *De memoria et reminiscentia. Memoria e reminiscenza*, edited by G. Nastasi, Ist. Italiano Studi Filosofici, Napoli 2024.

Aristotle, *On Sleep and Waking*, in *On Sleep and Dreams*, Engl. transl. by D. Gallop, Liverpool University Press, Liverpool 1996.

Aristotle, *Poetics*, Engl. transl. by M. Heath, Penguin, London 1997.

Aristotle, *Rhetoric*, Engl. transl. by G. A. Kennedy, Oxford University Press 2007.

Aristotle, *The Works of Aristotle: De Anima*, Engl. transl. by J.A. Smith, Clarendon Press, Oxford 1931.

Arnheim R., *Art and Visual Perception. A Psychology of the Creative Eye*, University of California Press, Berkeley and Los Angeles 1954.

Arnheim, R., *Visual Thinking*, University of California Press, Berkeley-Los Angeles 1969.

Asimov, I., *I, Robot*, Gnome Press, New York 1950.

Astolfi, A., *Phantasia in Aristotele*, Vita & Pensiero, Roma 2011.

Aubert, N., *Le culte de l'urgence. La société malade du temps*, Flammarion, Paris 2003.

Aumont, A., *L'image*, Armand Colin, Paris 2005.

Bachelard, G., *L'air et les songes. Essai sur l'imagination du mouvement*, Corti, Paris 1943.

Bachelard, G., *The Poetics of Reverie: Childhood, Language, and the Cosmos*, Engl. transl. by D.S. Russell, Beacon Press, Boston 1971.

Bachini, V. and Tesconi, M., *Fake people*, Codice Edizioni, Torino 2020.

Baczko, B., *Les imaginaires sociaux. Mémoirs et espoirs collectifs*, Payot, Paris 1984.

Baczko, B., headword *Immaginazione sociale*, in R. Romano (ed.), *Enciclopedia Einaudi*, vol. VII, Einaudi, Torino 1979.

Balàzs B., *Der sichtbare Mensch oder die Kultur des Films*, Deutsch-Österreichischer Verlag, Wien 1924.

Ball, M., *The Metaverse: And How It Will Revolutionize Everything*, WW Norton & Co, New York 2022.

Bajak, M. and Spendel, L., "The Possibilities of Using Artificial Intelligence in Advertising: Theory and Practice," in «Scientific Papers of Silesian University of Technology – Organisation and Management Series», 197, 2024, pp. 9-23.

Bansal, P., "Prompt Engineering Importance and Applicability with Generative AI," in «Journal of Computer and Communications», 12, 10, 2024.

Barale, A. (ed.), *Arte e intelligenza artificiale. Be my GAN*, Jaca Book, Milano 2020.

Barale, A., *Arte e intelligenza artificiale. Alcune domande*, in Ead. (ed.), *Arte e intelligenza artificiale. Be my GAN*, Jaca Book, Milano 2020, pp. 7-18.

Barthes, R., *Image Music Text*, Engl. transl. by S. Heath, Hill and Wang, New York 1977.

Baudrillard, J., *The Consumer Society. Myths and Structures* (1974), Sage Publications, London 1998.
Baudrillard, J., *The System of Objects*, Engl. transl. by J. Benedict, Verso, London 1996.
Bauman, Z., *Homo Consumens*, Erickson, Milan 2007.
Bauman, Z., *Identity. Conversations with Benedetto Vecchi*, Polity Press, London 2004.
Bauman, Z., *Liquid Times: Living in an Age of Uncertainty*, Polity Press, London 2000.
Baxandall, M., *Painting and Experience in Fifteenth Century Italy*, Clarendon Press, Oxford 1972.
Bayles, M. (ed.), *Contemporary Consequentialism*, Garden City, NY: Doubleday 1968.
Beauchamp, T.L. and Childress, J.F., *Principles of Biomedical Ethics* (1979), Oxford University Press, Oxford 2001.
Beck, U., *Conditio humana*, Laterza, Roma-Bari 2011.
Beck, U., *World Risk Society*, Polity Press, Cambridge 1999.
Belting, H., *An Anthropology of Images: Picture, Medium, Body*, Princeton University Press, Princeton & Oxford 2011.
Belting, H., "Image, Medium, Body: A New Approach to Iconology," in «Critical Inquiry», 31, 2, 2005, pp. 302-319.
Belting, H., *Likeness and Presence. A History of Image before the Era of Art*, The Chicago University Press, Chicago 1997.
Bender, E.M. *et al.*, "On the dangers of stochastic parrots: Can language models be too big?", in «Proceedings of the 2021 ACM Conference on Fairness, Accountability, and Transparency», 2021, pp. 610-2.
Benjamin, W., *The Work of Art in the Age of Its Technological Reproducibility* (1936, second version), in Id.,*The Work of Art in the Age of Its Technological Reproducibility, and Other Writings on Media*, Engl. transl. by M. W. Jennings *et al.* and eds. by E. Jephcott *et al.*, The Belknap Press of Harvard University Press, Cambridge Mass. London 2008.
Benjamin, W., *Walter Benjamin: Selected Writings, Volume 1 1913-1926*, edited by M. Bullock and M.W. Jennings, The Belknap Press of Harvard University Press, Cambridge Mass. – London 1996.
Bergamasco, M. *et al.*, *The Impact of Virtual and Augmented Reality on Individuals and Society*, Frontiers Media SA, Lausanne 2019.
Berlin, I., *Four Essays on Liberty*, Oxford University Press, Oxford 1969.
Berners-Lee, T., *Weaving the Web: The Original Design and Ultimate Destiny of the World Wide Web by Its Inventor*, Harper, 1999.

Bernstein, P.L., *Against the Gods. The Remarkable Story of Risk*, Wiley, New York 1996.
Bertolaso, M. and Marcos, A., *Umanesimo tecnologico*, Carocci, Roma 2024.
Bettetini, M., "The Status of Image: A Challenge for Medieval Aesthetics," in «Acta Philosophica», 2, 30, 2015, pp. 265-84.
Bianchi, F. *et al.*, "Easily Accessible Text-to-Image Generation Amplifies Demographic Stereotypes at Large Scale," in «FAccT '23: Proceedings of the 2023 ACM Conference on Fairness, Accountability, and Transparency», 2023, pp. 1493-1504.
Bodei, R., *Dominio e sottomissione. Schiavi, animali, macchine, Intelligenza Artificiale*, il Mulino, Bologna 2019.
Boden, M.A., *Artificial Intelligence: A Very Short Introduction*, Oxford University Press, Oxford 2018.
Boden, M.A., *Creativity and Art: Three Roads to Surprise*, Oxford University Press, Oxford/New York 2010.
Boden, M.A., *Creativity and Artificial Intelligence: A Contradiction in Terms*?, in E.S. Paul, S.B. Kaufman (eds.), *The Philosophy of Creativity: New Essays*, Oxford University Press, Oxford/New York 2014, pp. 224-44.
Boden, M.A., *The Creative Mind: Myths and Mechanisms*, Routledge, London 2004.
Boehm, G., *Die Wiederkehr der Bilder*, in Id. (ed.), *Was ist ein Bild?*, Fink, München 1994, pp. 11-38.
Boehm G. and Mitchell W.J.T., "Pictorial Versus Iconic Turn: Two Letters," in «Culture, Theory and Critique», 50, 2009, pp. 103-121.
Böhme, G., *Theorie des Bildes*, Fink, München 1999.
Bok, V., *Image Generation Models. GANs, Diffusion Models, and Transformers*, Manning Publications, New York 2025.
Bolin, G., (ed.), *Cultural Technologies: The Shaping of Culture in Media and Society*, Routledge, New York 2012.
Bolter, J.D. and Grusin, R., *Remediation: Understanding New Media*, MIT Press, Cambridge Mass. 2000.
Bondielli, A., *et al.*, "Mining Gender Biases in Text-to-Images Language Models," in «LaGendA Workshop onGender-inclusive language in a multilingual Europe», Venice, 3-4, October 2024.
Bordalo, P. *et al.*, "Stereotypes," in «Quarterly Journal of Economics», 131, 4, 2016, pp. 1753-94.
Bosco, H., *Un rameau de la nuit*, Flammarion, Paris 1950.
Bottom, W.P. and Kong, D.T., ""The Casual Cruelty of our Prejudices": On Walter Lippmann's Theory of Stereotype and Its "Obliteration"

in Psychology and Social Science," in «Journal of the History of the Behavioral Sciences», 48, 4, 2012, pp. 363-394.

Boulding, K.E., *Economics as a Science*, McGraw-Hill, New York 1970, p. 16.

Bourdieu, P., *Outline of a Theory of Practice*, Cambridge University Press, Cambridge 1977.

Bourdieu, P., *The Social Structures of the Economy*, Polity Press, Cambridge 2005.

Bovens, M., "Two concepts of accountability: accountability as a virtue and as a mechanism," in «West European Politics», 33, 5, 2010, pp. 946-7.

Brey, P. and Dainow, B., "Ethics by Design for Artificial Intelligence," in «AI Ethics», 4, 2024, pp. 1265-77.

Bruno, G., *Sigillus sigillorum* (1583), in Id., *Opera latine conscripta*, II, 2, edited by F. Fiorentino, F. Tocco, H. Vitelli, *et al.*, Neapoli Florentine 1879-1891.

Bruns, A., *Are Filter Bubbles Real?*, Polity Press, London 2019.

Bryn, D., *The Transparent Society*, Perseus Books, New York 1998.

Buber, M., *Das dialogische Prinzip* (1923), Engl. transl. by R.G. Smith, *I and You*, T. & T. Clark, Edinburgh 1938.

Buijsman, S. and Klenk, M., *AI, Ethics and Philosophy*, in N.A. Smuha (ed.), *The Cambridge Handbook of the Law, Ethics and Policy of Artificial Intelligence*, Cambridge University Press, Cambridge 2025, pp. 15-130.

Campbell, C., *AI by Design a Plan for Living with Artificial Intelligence*, Routledge, London 2022.

Canastrari, P. and Romeo, A. (eds.), *Second life. Oltre la realtà il virtuale*, Lupetti, Milano 2008.

Cao, Y. *et al.*, "A Comprehensive Survey of AI-Generated Content (AIGC): A History of Generative AI from GAN to ChatGPT", in «arXiv:2303.04226», March 2023.

Casetti, F., *Eye of the Century: Film, Experience, Modernity*, Columbia University Press, New York 2008.

Cassirer, E. *The Philosophy of the Enlightenment*, Engl. transl. by C.A. Koelln and J.P. Pettgrove, Princeton University Press, Princeton 1951.

Castelle, M., *La vita sociale delle reti antagoniste generative (GAN),* in A. Barale (ed.), *Arte e intelligenza artificiale. Be my GAN*, Jaca Book, Milano 2020.

Castoriadis, C., *L'institution imaginaire de la société*, Éditions du Seuil, Paris 1975, Engl. transl. by Blarney, K., *The Imaginary Institution of Society*, Polity Press, London 1987.

Catena, E. *et al.*, "Human autonomy with AI in the loop," in «Philosophical Psychology», 2025, pp.1-28.
Chakraborty, A. *et al.* (eds.), *Ethics in Artificial Intelligence: Bias, Fairness and Beyond*, Springer, Singapore 2023.
Clark, A. and Chalmers, D., "The Extended Mind," in «Analysis», 58, 1, 1998, pp. 7-19.
Chalmers, D.J., "Could a Large Language Model Be Conscious?," in «Boston Review» online, 2023, https://www.bostonreview.net/articles/could-a-large-language-model-be-conscious/.
Chalmers, D.J., "Facing up to the problem of consciousness," in «Journal of Consciousness Studies», 2,3, 1995, pp. 200-19.
Chalmers, D.J., *Reality+: Virtual Worlds and the Problems of Philosophy*, Allen Lane, London 2022
Chalmers, D.J., *The Conscious Mind*, Oxford University Press, Oxford 1996.
Chatonsky, G., "Deep Dream (The Network's Dream)," in
Cheney-Lippold, J., *We Are Data: Algorithms and the Making of Our Digital Selves*, New York University Press, New York 2018.
Chiriatti, M., *Incoscienza artificiale*, Luiss University Press, Roma 2022.
Chock, S.C., *Design Justice*, MIT Press, Cambridge Mass. 2020.
Chun, J. *et al.*, "Comparative Global AI Regulation: Policy Perspectives from the EU, China, and the US", in «arXiv:2410.21279» 2024.
Clapp, J., "The Ethics of Alterity and the Ethics of Care in Literary Studies," in «Educational Philosophy and Theory», 57, 3, 2024, pp. 262-73.
Clouser, K.D. and Gert, B., "A Critique of Principlism," in «Journal of Medicine and Philosophy», 15, 2, 1990, pp. 219-236.
Coeckelberg, M., *AI Ethics*, MIT Press, Cambridge Mass.- London 2020.
Conti, M. and Tricomi, P.P., "Virtual Influencers in Online Social Media", in «IEEE Communications Magazine», 60, 2022, pp. 1-13.
Corbin, H., "Mundus Imaginalis, or the Imaginary and the Imaginal," in «Cahiers internationaux de symbolisme», 6, 1964, pp. 3-26.
Correa, N.K. *et al.*, "Worldwide AI ethics: A review of 200 guidelines and recommendations for AI governance," in «Patterns», 4, 202279.
Crawford, K., *Atlas of AI: Power, Politics, and the Planetary Costs of Artificial Intelligence*, Yale University Press, New Haven Conn. 2021.
Crawford, K. and Paglen, T., "Excavating AI: the Politics of Images in Machine Learning Training Sets," in «AI & Society», 36, 4, 2021, p. 1399.
Crary, J., *Techniques of the Observer. On Vision and Modernity in the Nineteenth Century*, MIT Press, Cambridge Mass. 1990.
Cristianini, N., *Machina Sapiens*, il Mulino, Bologna 2024.

Cristianini, N., *The Shortcut. Why Intelligent Machines Do Not Think Like Us*, Routledge, London 2023.

Custers, B. and Fosch-Villaronga, E., *Law and Artificial Intelligence. Regulating AI and Applying AI in Legal Practice*, Springer, Berlin 2022.

Dadà, S., *Vulnerabilità digitale: etica, intelligenza artificiale e medicina*, Mimesis, Milano 2024.

Dalferth I., U. and Perrier, R.E. (eds.), *Humanity: An Endangered Idea? Claremont Studies in the Philosophy of Religion*, Mohr Siebeck, Tübingen 2023, pp. 351-362

Daskas, B., *La lunga vita dell'ekphrasis tra Bisanzio e la contemporaneità*, in F. Conca and C. Castelli (eds.), *Bisanzio fra tradizione e modernità Ricordando Gianfranco Fiaccadori*, Ledizioni, Milano 2017.

Daub, A., *The Cancel Culture Panic: How an American Obsession Went Global*, Stanford University Press, Stanford 2024.

Dean, M. "Risk, Calculable and Incalculable," in «Soziale Welt», 49, 1998, pp. 25-42.

Debord, G., *La société du spectacle*, Gallimard, Paris 1992, Engl. transl. by K. Knabb, *The Society of the Spectacle*, Bureau of Public Secrets, Berkeley CA 2014.

Debray, R., *Vie et mort de l'image*, Éditions Gallimard, Paris 1992.

Debray, R., *Media Manifestos*, Engl. transl. by E. Rauth, Verso, London 1996.

De Kerckhove, D., *The Skin of Culture: Investigating the New Electronic Reality*, Kogan Page, London 1998.

De la Peña, N. *et al.*, "Immersive Journalism: Immersive Virtual Reality for the First-Person Experience of News," in *«Presence: Teleoperators and Virtual Environments»*, 19, 4, 2010, pp. 291-301.

Deleuze, G., *Postscript on Control Societies* (1990), in Id., *Negotiations*, Engl. transl. by M. Joughin, Columbia University Press, New York 1985, pp. 177-82.

Denson, S., "Artificial Imagination," in «Cinephile: The University of British Columbia's Film Journal», 18, 1, 2024.

Derrida, J., *Répondre – du secret, Volume I, Séminaire (1991-1992) – Secret et témoignage,* ed. by G. Michaud and N. Cotton, Suil, Paris 2024.

Descartes, R., *Œuvres complètes*, eds. M. Beyssade and D. Kambouchner, Gallimard, Paris 2018.

Devine, P.G., "Stereotypes and prejudice: Their automatic and controlled components," in «Journal of Personality and Social Psychology», 56, 1, 1989, pp. 5-18.

Devine, P.G. and Sharp, L.B., *Automaticity and Control in Stereotyping and Prejudice*, in T. D. Nelson (ed.), *Handbook of Prejudice Stereotyping, and Discrimination*, American Psychological Association, Psychology Press, Hove 2009, pp. 61-87.

Diderot, D., and D'Alambert, J., *L'Encyclopédie*, 1re éd., David l'Aîné Briasson, Durand Le Breton, Paris 1766, Tome 8, pp. 560-564.

Diderot, D., and D'Alambert, J., *Immaginazione*, in D. Diderot *et al.*, *L'estetica dell'Encyclopédie*, edited by M. Modica, Editori Riuniti, Roma 1995.

Dignum, V., *et al.*, "Ethics by Design: Necessity or Curse?" in «Proceedings of the 2018 AAAI/ACM Conference on AI, Ethics, and Society», 2018, pp. 60-66.

Di Monte, M., "Iconosphere," in «International Lexicon of Aesthetics», 1, 2018.

Diodato, R., *L'arte come categoria estetica. Un'introduzione*, Eupress FTL, Lugano 2005.

Di Pasquale, F., *The Black Box Society: The Secret Algorithms That Control Money and Information*, Harvard University Press, Cambridge Mass. 2016.

D'Isa, F., *La rivoluzione algoritmica delle immagini. Arte e intelligenza artificiale*, Luca Sossella Editore, Milano 2024.

Doerner, R. and Broll, W. (eds.), *Virtual and Augmented Reality (VR/AR): Foundations and Methods of Extended Realities (XR)*, Springer Nature Switzerland AG 2022.

Donise, A., *Critica della ragione empatica. Fenomenologia dell'altruismo e della crudeltà*, il Mulino, Bologna 2020.

Douglas, M., *Risk and Blame. Essays in Cultural Theory*, Routledge, London 1992.

Douglas, M. and Wildavsky, A., *Risk and Culture*, University of California Press, Berkeley 1982.

Dowling, K., *et al.*, "Behavioral Biases in Marketing," in «Academy of Marketing Science», 48, 2020, pp. 449-77.

Drenten, J. and Brooks, G., "Celebrity 2.0: Lil Miquela and the rise of a virtual star system", in «Feminist Media Studies», 20, 8, 2020, pp. 1319-23.

Dreyfus, H. and S. Dreyfus, S., *What Computers Still Can't Do: A Critique of Artificial Reason*, MIT Press, Cambridge Mass. 1992.

Durand, G., *L'imagination symbolique* (1964), PUF, Paris 1984.

Dworkin, R., *Taking Rights Seriously*, Harvard University Press, Harvard 1977.

Ellington, J.W., *Grounding for the Metaphysics of Morals*, Hackett Publishing Company, Indianapolis 1993.

Esposito, E., *Artificial Communication. How Algorithms Produce Social Intelligence*, MIT Press, Cambridge Mass. 2022.

Fabris, A., *Aesthetics and Ethics in Technological Environments*, in Ž. Paić (ed.), *The Technosphere as a New Aestetic*, Cambridge Scholar Publishing, NewCastle 2022, pp. 74-88.

Fabris, A., *Ethics of Information and Communication Technologies*, Springer, Cham 2018.

Fabris, A., *Etica del giornalismo*, Edizioni ETS, Pisa 2019.

Fabris, A., *Etica della comunicazione*, Carocci, Roma 2014.

Fabris, A., *Premessa*, in A. Fabris (ed.), *Etiche applicate. Una guida*, Carocci, Roma 2018, pp. 11-8

Fabris, A., *La filosofia nell'epoca dell'intelligenza artificiale. Come ci aiuta a vivere in ambienti tecnologici*, Carocci, Roma 2025.

Fabris, A., *La scelta del dialogo. Breviario filosofico per comunicare meglio*, Messaggero, Padova 2011.

Fabris, A. and Ambriola,V., "Per un'etica dell'intelligenza artificiale nei sistemi socio-tecnologici," in «Estetica. Studi e ricerche», 2, 2024, pp. 339-54.

Fabris, A., *Philosophy, Image and the Mirror of Machines*, in Ž. Paić and K. Purgar (eds.), *Theorizing Images*, Cambridge Scolars Publishing, Newcastle 2016, pp. 111-20.

Fabris, A., *RelAzione. Una filosofia performativa*, Morcelliana, Brescia 2016.

Fabris, A., and Belardinelli, S. (eds.), *Digital Environments and Human Relations. Human Perspectives in Health Sciences and Technology*, vol. 150, Springer, Cham 2024.

Fabris, A. *et al.*, *Towards a Relational Ethics in AI. The Problem of Agency, The Search for Common Principles, the Pairing of Human and Artificial Agents*, in A. Fabris, S. Belardinelli (eds), *Digital Environments and Human Relations. Human Perspectives in Health Sciences and Technology*, vol 150, Springer, Cham 2024, pp. 9-42.

Fabris A., *The Transformation of Human Ethics in an Age of Technological Challenges*, in Dalferth I.U. and Perrier R.E. (eds.), *Humanity: An Endangered Idea? Claremont Studies in the Philosophy of Religion*, Mohr Siebeck, Tübingen 2023, pp. 351-362.

Falcinelli, R., *Figure. Come funzionano le immagini dal Rinascimento a Instagram*, Torino, Einaudi 2020.

Falcinelli, R., *Visus*, Einaudi, Torino 2024.

Farisco, M., Evers, K., Changeux, J.P., "Is Artificial Consciousness Achievable? Lessons from the Human Brain," in «Neural Networks», 180, 106714, 2024.

Ferrara, E., "Fairness and Bias in Artificial Intelligence: A Brief Survey of Sources, Impacts, and Mitigation Strategies," in «Sci», 6 1, 3, 2023.
Ferrarin, A., *Un mondo non di questo mondo*, Edizioni ETS, Pisa 2023.
Ferraris, M., *L'immaginazione*, Il Mulino, Bologna 1996.
Feyles, M., "Kant e la fenomenologia dell'immaginazione", in « Aesthetica Preprint», 120, 2022, pp. 65-78.
Feuerriegel, S., *et al.*, "Generative AI," in «Business & Information Systems Engineering», 66, 2024, pp. 111-126.
Filimowicz, M., *Systemic Bias: Algorithms and Society*, Routledge, London 2023.
Finn, E., *What Algorithms Want. Imagination in the Age of Computing*, MIT Press, Cambridge Mass 2017.
Fischer, J.M. and Tognazzini, N., "The Physiognomy of Responsibility," in «Philosophy and Phenomenological Research», 82, 2, 2011, pp. 381-417.
Fiske, S.T. and Taylor, S.E., *Social Cognition: From brains to culture* (1984), Random House, New York 1988.
Fiske, S.T. *et al.*, "A model of (often mixed) stereotype content: Competence and warmth respectively follow from perceived status and competition," in «Journal of Personality and Social Psychology», 82, 6, 2002, pp. 878-902.
Floridi, L., "Ethics in the Infosphere," in «The Philosophers' Magazine», 16, 2001, pp. 18-9.
Floridi, L., *Ethics of Artificial Intelligence. Principles, Challenges, and Opportunities*, Oxford University Press, Oxford 2023.
Floridi, L., "Four challenges for a theory of informational privacy," in «Ethics and Information Technology», 8, 2006, pp.109-19.
Floridi, L., *Pensare Infosfera. La filosofia come design concettuale*, Raffaello Cortina Editore, Milano 2020.
Floridi, L., *The Ethics of Information*, Oxford University Press, Oxford 2013.
Floridi, L., *The Fourth Revolution: How the Infosphere is Reshaping Human Reality*, Oxford University Press, Oxford 2014.
Floridi, L., *The Logic of Information: A Theory of Philosophy as Conceptual Design*, Oxford University Press, Oxford 2019.
Floridi, L., *The Onlife Manifesto: Being Human in a Hyperconnected Era*, Springer, Cham 2014.
Floridi, L., *The Philosophy of Information*, Oxford University Press, New York 2011.
Floridi, L. and Cowls, J., "A Unified Framework of Five Principles for AI in Society", in «Harvard Data Science Review», 1, 1, 2019.

Filimowicz, M., *Deep Fakes: Algorithms and Society*, Routledge, London 2022.

Finocchiaro, G., *Intelligenza artificiale. Quali regole*, Il Mulino, Bologna 2024.

Flusser, V., *Bilderstatus*, in, *Medienkultur*, ed. by S. Bollmann, Fischer Taschenbuch, Frankfurt am Main 1997, pp. 69-82.

Flusser, V., *Die kodifizierte Welt*, in Id., *Medienkultur*, ed. by S. Bollmann, Fischer Taschenbuch, Frankfurt am Main 1997, pp. 21-28.

Flusser, V., *Eine neue Einbildungskraft*, in Id., *Bildlichkeit*, Suhrkamp, Frankfurt am Main 1990, pp. 115-26.

Flusser, V., *Filmerzeugung und Filmverbrauch*, in Id., *Medienkultur*, ed. by S. Bollmann, Fischer Taschenbuch, Frankfurt am Main 1997, pp.89-102.

Flusser, V., *Für eine Philosophie der Fotografie*, European Photography, Göttingen 1983, Engl. transl. by A. Mathews, *Towards a Philosophy of Photography*, Reaktion Books, London 2000.

Flusser, V., *Gesten: Versuch einer Phänomenologie*, Bollmann, Düsseldorf 1991.

Flusser, V., *Heinweg von Papier*, in Id., *Medienkultur,* ed. by S. Bollmann, Fischer Taschenbuch, Frankfurt am Main 1997, pp. 61-66.

Flusser, V., *Ins Universum der technischen Bilder*, European Photography, Göttingen 1985, Engl. transl. by N.A. Roth, *Into the Universe of Technical Images*, University of Minnesota Press, Minneapolis 2011.

Flusser, V., *Kommunikologie*, hrsg. Von S. Bollmann und E. Flusser, Frankfurt a.M. 1998.

Flusser, V., *Medienkultur*, ed. by S. Bollmann, Fischer Taschenbuch, Frankfurt am Main 1997.

Flusser, V., *Nomadische Überlegungen*, in Id., *Medienkultur,* ed. by S. Bollmann, Fischer Taschenbuch, Frankfurt am Main 1997, pp. 150-159.

Flusser, V., *O Último Juízo*, vol. II, È Realizações Editora, São Paulo 2017.

Franco, V., *Responsabilità. Figure e metamorfosi di un concetto*, Donzelli, Milano 2015.

Frankfurt, H.G., "Alternate Possibilities and Moral Responsibility," in «The Journal of Philosophy», 66, 23, 1969, pp. 829-39.

Frankfurt, H.G., "Freedom of the Will and the Concept of a Person," in «The Journal of Philosophy», 67, 1, 1971, pp. 5-20.

Frankish, K., "Dual-process and Dual-system Theories of Reasoning," in «Philosophy Compass», 5, 10, 2010, pp. 914-26.

Franzini, E., *Un'immaginazione per il Settecento*, in F. Carmagnola and V. Matera (eds.), *Genealogie dell'immaginario*, UTET, Torino 2008.

Freeberg, D., *The Power of Images: Studies in the History and Theory*, The University of Chicago Press, Chicago 1989.

Frey, R.G., *A Companion to Applied Ethics*, Blackwell, Oxford 2004.
Friedman, M., *Martin Buber's Life and Work: the Early Years*, 1878-1923, Dutton, New York 1982.
Friedman, B. and Hendry, D.G., *Value Sensitive Design. Shaping Technology with Moral Imagination*, MIT Press, Cambridge Mass. 2019.
Fui-Hoon Nah F. *et al.*, "Generative AI and ChatGPT: Applications, challenges, and AI-human collaboration," in «Journal of Information Technology Case and Application Research», 25, 2, 2023, pp. 289-290.
Gadamer, H.G., *Wort und Bild – 'so wahr, so seiend'* (1992), in Id., *Gesammelte Werke*, 10 vols., Tübingen, Mohr, 1985-1995, vol. 8 (1993), pp. 373-99.
Gaur, L. (eds.), *Deepfake. Creation, Detection and Impact*, CRC Press, Boca Raton 2022.
Geiger, M., *Sull'essenza e il significato dell'empatia*, in A. Pinotti (ed.), *Estetica ed empatia*, Guerini, Milano 1997, pp. 61-94.
Gell, A., *Art and Agency: An Anthropological Theory*, Clarendon Press, Oxford 1998.
Ghosh, S. and Caliskan, A., *'Person' == Light-skinned, Western Man, and Sexualization of Women of Color: Stereotypes in Stable Diffusion*, in H. Bouamor *et al.*(eds.), *Findings of the Association for Computational Linguistics: EMNLP 2023*, Association for Computational Linguistics, Singapore 2023, pp. 6971-85.
Giannotti, F., Naretto F. and Bodria, F., *Explainable for Trustworthy AI*, in Chetouani M. et al. (eds.), *Human-Centered Artificial Intelligence Advanced Lectures*, Springer Science and Business Media Deutschland GmbH, Berlin-Heidelberg 2023, pp. 175-195.
Giddens, A., *The Consequences of Modernity*, Stanford University Press, Stanford 1990.
Gigerenzer, G., *Gut Feelings: The Intelligence of the Unconscious*, Penguin Publishing Group, New York 2007.
Ginger, S., *Gestalt Therapy: The Art of Contact*, Routledge, London 2007.
Giovanola, B., Tiribelli,S., "Beyond bias and discrimination: redefining the AI ethics principle of fairness in healthcare machine-learning algorithms", in «AI & Society», 38, 2023, pp. 549-563.
Girard, R., *Je vois Satan tomber comme l'éclair*, Grasset, Paris 1999.
Girard, R., *The Scapegoat* (1982), Johns Hopkins University Press, Baltimore 1999.
Glaser, H. and Wong, P., *Governing the Future: Digitalization, Artificial Intelligence, Dataism,* CRC Press, Boca Raton 2025.
Gombrich, E.H., *Art and Illusion: A Study in the Psychology of Pictorial Representation*, Pantheon Books, New York 1960.

Gombrich, E.H., *The Sense of Order: A Study in the Psychology of Decorative Art*, Cornell University Press, Ithaka & New York 1984.

Goodfellow J., *et al.*, *Deep Learning*, MIT Press, Cambridge Mass. 2016.

Goodfellow, J. *et al.*, *Generative Adversarial Networks*, in «ArXiv:1406.2661v1», June 2014.

Gunkel, D.G., *The Machine Question: Critical Perspectives on AI, Robots, and Ethics*, MIT Press, Cambridge Mass. 2012.

Hagendorff, T., "The Ethics of AI Ethics: An Evaluation of Guidelines", in «Minds and Machines», 30, 2020, pp. 99-120.

Han, B.-C., *Good Entertainment: A Deconstruction of the Western Passion Narrative*, Engl. transl. by A.N. West, MIT Press, Cambridge Mass 2019.

Han, B.-C., *Infocracy. Digitisation and the Crisis of Democracy*, Engl. transl. by D. Steuer, Polity Press, Cambridge 2022.

Han, B.-C., *Non-things. Upheaval in the Lifeworld*, Engl. transl. by D. Staeler, Polity Press, London 2022.

Han, B.-C., *Psychopolitics: Neoliberalism and New Technologies of Power*, Verso, New York 2017.

Han, B.-C., *The Crisis of Narration*, Engl. transl. by D. Steuer, Polity Press, London 2024.

Han, B.-C., *The Palliative Society: Pain Today*, Engl. transl. by D. Steuer, Polity Press, London 2021.

Han, B.-C., *The Transparency Society*, Stanford University Press, Stanford 2015.

Harari, Y. N., Homo Deus, *A Brief History of Tomorrow* (2016), Harper Collins, New York 2017.

Harari, Y. N., *Nexus. A Brief History of the Information Network from the Stone Age to AI*, Vintage Digital Publishing 2024.

Härle, Cl.-C. (ed.), *Ai limiti dell'immagine*, Quodlibet, Macerata 2005.

Hart, H.L.A., *Punishment and Responsibility*, Oxford University Press, Oxford 1968.

Hegel, G.W.F., *Aesthetics* (1835), by M. Knnx, Clarendon Press, Oxford 1975.

Heidegger, M., *The Age of the World Picture* (1938), in Id., *Off The Beaten Track*, Engl. ed. and transl. by J. Yung K. Haynes, Cambridge University Press, Cambridge 2002, pp. 57-84.

Helberger, N., "Choice Architectures in the Digital Economy: Towards a New Understanding of Digital Vulnerability," in «Journal Consum Policy», 45, 2, 2021, pp. 175-200.

Henin, S., *Il computer dimenticato. Charles Babbage, Ada Lovelace e la ricerca della macchina perfetta*, edited by M. Temporelli, Hoepli, Milano 2015.

Hillman, J., *The Practice of Beauty*, in B. Beckley and D. Shapiro D. (Eds.), *Uncontrollable Beauty, toward a New Aesthetics*, Allworth Press, New York, 1998, pp. 261-274.

Hobbes, T., *Leviathan, or the Matter, Form and Power of an Ecclesiastical and Civil Commonwealth* (1651), edited by C. Brooke, Penguin Classics, London 2017.

Hochschild, A.R., *The Commercialization of Intimate Life*, University of California Press, Berkeley 2003.

Hoelzl, I. and Marie, M., *Postimage*, in R. Braidotti and M. Hlavajova (eds.), *Posthuman Glossary*, Bloomsbury Academic, London-New York 2018, pp. 360-362.

Hoffman, M.L., *Empathy and Moral Development: Implications for Caring and Justice*, Cambridge University Press, Cambridge 2000.

Hume, D., *A Treatise of Human Nature* (1738), ed. by L.A. Selby Bigge, Clarendon Press, Oxford 1896.

Humphreys, P., "The Philosophical Novelty of Computer Simulation Methods," in «Synthese», 169, 3, 2008, pp. 615-26.

Hund, E., *The Influencer Industry: The Quest for Authenticity on Social Media,* Princeton University Press, Princeton 2023.

Husserl, E., *Ideen zu einer reinen Phänomenologie und phänomenologischen Philosophie*, Niemeyler, Halle 1913, Engl. transl. by F. Kersten, R. Rojcewicz and A. Schuwer, *Ideas Pertaining to a Pure Phenomenology and to a Phenomenological Philosophy*, Kluwer Academic Publishers, Dordrecht 1982.

Hustinx, P., "Privacy by Design: Delivering the Promises," in «Identity in the Information Society», 3, 2, 2010, pp. 253-5.

Irrgang, D., *Vilém Flussers Black Box*, in E. Geitz, C. Vater, S. Zimmer-Merkle (eds.), *Interdisziplinare Perspektiven*, De Gruyter, Berlin-Boston 2020, pp. 53-69.

Jazdzewska, K., "From Dialogos to Dialogue: The Use of the Term from Plato to the Second Century CE," in «Greek, Roman and Byzantine Studies», 54,1, 2015, pp. 17-36.

Jenks, C. (eds.), *Visual Culture*, London, Routledge 1995.

Jobin, A. *et al.*, "Artificial Intelligence: the global landscape of ethics guidelines," in «Health Ethics & Policy Lab», 2019.

Johnson, A.T., *Consciousness for Artificial Intelligence?,* 2024, https://www.embs.org/pulse/articles/consciousness-for-artificial-intelligence/.

Jonas, H., *The Imperative of Responsibility: In Search of an Ethics for the Technological Age*, Engl. transl. by H. Jonas, D. Herr, The University of Chicago Press, Chigaco-London 1984, Or. ed. *Das Prinzip Verantwortung : Versuch einer Ethik fur die technologische Zivilisation*, Insel Verlag Frankfurt am Main, 1979.

Jovanović, M. and Campbell, M., "Generative Artificial Intelligence: Trends and Prospects," in «Computer», October, 2022, pp. 107-112.

Kagan, S., *Normative Ethics*, Routledge, London 1997.

Kahneman, D., *Thinking, Fast and Slow* (2011), Farrar, Straus & Giroud, New York 2013.

Kahneman, D. and Frederick, S., *Heuristics and Biases: The Psychology of Intuitive Judgment*, Cambridge University Press, New York 2002.

Kahneman, D., *et al.*, *Judgment under Uncertainty. Heuristics and Biases*, Cambridge University Press, Cambridge 1982.

Kant, I., *Kritik der Reinen Vernunft*, in „*Kant's Gesammelte Schriften, hrsg. Von der Königlich preußischen Akademie der Wissenschaften*", Band III, Berlin 1904, Engl. transl. and edited by P. Guyer and A. Wood, *Critique of Pure Reason* 1781/1787, Cambridge University Press, Cambridge 1997.

Kant, I., *Kritik der Urteilskraft* (1790), in „*Kant's Gesammelte Schriften*", hrsg. Von der Königlich preußischen Akademie der Wissenschaften, Band V., Berlin 1908, ed. and transl. by P. Guyer, transl. by E. Matthews, *Critique of the Power of Judgment*, Cambridge University Press, Cambridge 2000.

Kant, I., *Grundlegung zur Metaphysik der Sitten* (1785), in *Kants Gesammelte Schriften*, hrsg. Königlichen Preußischen Akademie der Wissenschaften, Reimer, Berlin, vol. 4, 1903, Engl. transl. by J.W. Ellington, *Grounding for the Metaphysics of Morals*, Hackett Publishing Company, Indianapolis, 1993.

Kant, T., *Making it Personal: Algorithmic Personalization, Identity, and Everyday Life*, Oxford University Press, Oxford 2020.

Kaplan, M. and Haenlein, M., "Siri, Siri, in My Hand: Who's the Fairest in the Land? On the Interpretations, Illustrations, and Implications of Artificial Intelligence," in «Business Horizons», 62, 1, 2019, pp. 15-25.

Kardiner, A., Ovesey, L., *The Mark of Oppression: A Psychosocial Study of the American Negro*, W.W. Norton and Company, New York 1951.

Kaufman, S.B., (eds.), *The Philosophy of Creativity: New Essays*, Oxford University Press, Oxford/New York 2014, pp. 224-44.

Katz, D. and Braly, K., "Racial Stereotypes in One Hundred College Student," in «Journal of Abnormal and Social Psychology», 28, 1933, pp. 280-90.

Kiener, M. (ed.), *The Routledge Handbook of Philosophy of Responsibility*, Routledge, London 2024.

Kinsey, C., *Walled Gardens: Autonomy, Automation, and Art After the Internet*, Oxford University Press, Oxford 2021.

Kirova, V. *et al.*, "The Ethics of Artificial Intelligence in the Era of Generative AI," in «Journal of Systemics Cybernetics and Informatics», 21, 4, 2023, pp. 42-50.

Kleingeld, P., *The Principle of Autonomy in Kant's Moral Theory: Its Rise and Fall*, in E. Watkins (ed.), *Kant on Persons and Agency*, Cambridge University Press, Cambridge 2017, pp. 61-80.

Klingemann, M., *Memories of Passersby I*, in A. Barale (ed.), *Arte e intelligenza artificiale. Be my GAN*, Jaca Book, Milano 2020, pp. 75-88.

Klingemann, M., *Uncanny Mirror*, in A. Barale (ed.), *Arte e intelligenza artificiale. Be my GAN*, Jaca Book, Milano 2020, pp. 97-102.

Köhler, W., *Gestalt Psychology: An Introduction to New Concepts in Modern Psychology*, Liveright Publishing Corporation, New York 1992.

Kulkarny, C. *e al.*, *A Word is Worth a Thousand Pictures: Prompts as AI Design Material*, in «arXiv:2303.12647», 2023.

Kunda, Z., *Social Cognition: Making Sense of People*, MIT University Press, Cambridge Mass. 1999.

Lagioia, F. and Sartor, G., "Profilazione e decisione algoritmica: dal mercato alla sfera pubblica," in «Federalismi- Focus: La disinformazione online e il ruolo degli esperti nell'agorà digitale», 11, 2020, pp. 85-110.

Leibniz, G.W., *New Essays on the Human Intellect* (1765), It. transl. by D.O. Bianca, *Nuovi saggi sull'intelletto umano*, Utet, Torino 1976.

Lipps, T., *Ästhetik*, Voss Verlag, Hamburg 1903-06.

Luccioni, A.S. *et al.*, "Stable Bias: Analyzing Societal Representations in Diffusion Models," in «Science Computers and Society», preprint arXiv:2303.11408, 2023.

Luccioni, A.S., "Stable Bias: Evaluating Societal Representations in Diffusion Models," in «NIPS '23: Proceedings of the 37th International Conference on Neural Information Processing Systems», 2458, pp. 56338-56351

Luhmann, N., *The Reality of the Mass Media*, Engl. transl. by K.Cross, Stanford University Press, Standford 2000.

Luhmann, N., *Generalized Media and the Problem of Contingency*, in J.J. Loubser *et al.* (eds.), *Exploration in General Theory of Social Science*, Free Press, New York 1976, pp. 506-532.

Luhmann, N., *Risk. A Sociological Theory*, Aldine de Gruyter, New York 1993.

Malebranche, N., *De la recherche de la vérité* (1674), in Id., *Œuvres complètes*, edited by G. Rodis-Lewis, vol. II, Vrin, Paris 1962.

Manovich, L., *AI Aesthetics*, Strelka Press, Moscow 2018.

Manovich, L., and Aurielli, E., *Artificial Aesthetics: Generative AI, Art and Visual Media*, 2024, https://manovich.net/index.php/projects/artificial-aesthetics.

Manu, A., *Transcending Imagination. Artificial Intelligence and the Future of Creativity*, Routledge, London 2024.

March, J.G., *Primer on Decision Making: How Decisions Happen*, Free Press, Detroit 1994.

Marcuse, H., *One-Dimensional Man*, The Beacon Press, Boston 1964.

Marinucci, L. *et al.*, "Exposing implicit biases and stereotypes in human and artificial intelligence: state of the art and challenges with a focus on gender," in «AI & Society», 38, 2023, pp. 747-61.

Mayer-Schonberger, V. and Cukier, K., *Big Data: A Revolution That Will Transform How We Live, Work and Think*, John Murray Publishers Ltd, London 2013.

McCarthy, J. *et al.*, "A Proposal for the Dartmouth Summer Research Project on Artificial Intelligence, August 31, 1955, in «AI Magazine», 27, 4, 2006, pp. 12-14.

McCulloch, W.S. and Pitts, W., "A logical calculus of the ideas immanent in nervous activity," in «Bulletin of Mathematical Biophysics», 5, 1943, pp. 115-33.

McKinstry, C. "Minimum Intelligent Signal Test: An Alternative Turing Test," in «Canadian Artificial Intelligence», 41, 1997, 35-47.

McLuhan, M., *Understanding Media*, McGraw-Hill, New York 1964.

Menon, M. *Vilém Flusser e la «rivoluzione dell'informazione». Comunicazione, etica, politica*, Edizioni ETS, Pisa 2022.

Meskys, E. *et al.*, "Regulating Deepfakes: Legal and Ethical Considerations", in «Journal of Intellectual Property Law & Practice», 15, 2020, pp. 24-31.

Messinetti, R., *Comunicare nell'infosfera. La vulnerabilità della persona digitale*, 2021, in «Federalismi online»: https://www.federalismi.it/nv14/editoriale.cfm?eid=603.

Miano, F., *Responsabilità*, Orthotes, Napoli 2010.

Milano, S. *et al.*, "Recommender Systems and Their Ethical Challenges", in «AI & Society», 35, 2020, pp. 957-67.

Mirzoeff, N., *An Introduction to Visual Culture*, Routledge, London 1999.

Mitchell, M., *Artificial Intelligence: A Guide for Thinking Humans*, Farrar, Straus and Giroux 2019.

Mitchell, W.J.T., *Picture Theory: Essays on Verbal and Visual Representation*, University of Chicago Press, Chicago 1994.
Mitchell, W.J.T., *Pictorial Turn: eine Antwort*, in H. Belting (ed.), *Bilderfragen. Die Bildwissenschaften im Aufbruch*, Fink, München 2002, pp. 37-48.
Mitchell, W.J.T., "Pictorial turn. Una risposta," in «Lebenswelt», 2, 2012, pp. 130-42.
Mitchell, W.J.T., *The Reconfigured Eye Visual Truth in the Post-Photographic Era*, MIT Press, Cambridge Mass. 1994.
Mitchell, W.J.T., "What Do Pictures Really Want?" in «JSTOR», 77, 1996, pp. 71-82.
Mitchell, W.J.T., *What Do Pictures Want? The Lives and Loves of Images*, The University of Chicago Press, Chicago 2005.
Mittelstadt, B.D. *et al.*, "The Ethics of Algorithms: Mapping the Debate," in «Big Data & Society», 3, 2, 2016, pp. 1-21.
Mökander, J. and Floridi, L. "Ethics-Based Auditing to Develop Trustworthy AI," in «Minds & Machines», 31, 2021, pp. 323-327.
Mökander, J. *et al.*, "Ethics-Based Auditing of Automated Decision-Making Systems: Nature, Scope, and Limitations," in «Science and Engineering Ethics», 27, 4, 2021.
Morley, J. *et al.*, "Ethics as a Service: A Pragmatic Operationalisation of AI Ethics," in «Minds & Machines», 31, 2, 2021, pp. 239-56.
Mazzone, M., *Le GAN e la questione della creatività nell'arte e nell'intelligenza artificiale*, in A. Barale (ed.), *Arte e intelligenza artificiale. Be my GAN*, Jaca Book, Milano 2020, pp. 51-74.
Morris, G.S. (ed.), *Hegel's Aesthetics*, S.C. Criggs & Company, Chicago 1885.
Morton, J.B. *et al.*, "Age-related Changes in Brain Activation Associated with Dimensional Shifts of Attention: an fMRI Study," in «Neuroimage», 46, 1, 2009, pp. 249-56.
Moura, L., "Robot Art: An Interview with Leonel Moura," in «Arts», 7, 3, 2018.
Müller, V.C., *Philosophy of AI: A Structured Overview*, in N.A. Smuha (ed.), *The Cambridge Handbook of the Law, Ethics and Policy of Artificial Intelligence*, Cambridge University Press, Cambridge 2025, pp. 40-58.
Munari, B., *Da cosa nasce cosa*, Laterza, Roma-Bari 1981.
Munari, B., *Fantasia*, Laterza, Bari-Roma 1977.
Napolitano Valditara, L.M., *Platone e le "ragioni" dell'immagine: percorsi filosofici e deviazioni tra metafore e miti*, Vita & Pensiero, Milano 2007.

Neri, V., “Generative Artificial Intelligence, Deepfakes and Vulnerable Identity. The Ethics of Uncertainty as a Response to (In)controllable Risk,” in «Teoria», 2, Edizioni ETS, Pisa, 2024, pp. 79-94.

Neri, V., *L'immagine nel web. Etica e ontologia*, Carocci, Roma 2013.

Neri, V., *Notizie visive*, Pacini edizioni, Pisa 2021.

Neri, V., “The Autonomy and Responsibilities of the Image in Cyberspace”, in «Phainomena», 100/101, 2017, pp. 223-239.

Newman, L.S., “Was Walter Lippmann interested in Stereotyping?: Public Opinion and Cognitive Social Psychology,” in «History of Psychology», 12, 1, 2009.

Ng, E., *Cancel Culture. A Critical Analysis*, Palgrave, London 2022.

Nguyen, C.T., “Echo Chambers and Epistemic Bubbles,” in «Episteme», 17, 2, 2020, pp. 141-61.

Norton, B., “The Extinction Image Free”, in «Cultural Politics», 19, 3, 2023, pp. 333-52.

Novelli, C. Taddeo, M. and Floridi, L., “Accountability in Artificial Intelligence: What It Is and How It Works,” in «AI & Society», 39, 2024.

Novelli, C. *et al.,* “Taking AI Risks Seriously: A New Assessment Model For the AI Act,” in «AI & Society», 39, 2024, pp. 2493-97.

Nussbaum, M. and Rorty, A.M. (eds.), *Essays on Aristotle's De anima*, Clarendon Press, Oxford 1992.

Nussbaum, M., *Women and Human Development. The Capabilities Approach* (1999), Cambridge University Press, Cambridge 2001.

Ochoa, B., *AI-powered Advertising: Examining the Possibilities of AI in Creating Advertisements for All Persons*, Independent publishing, 2024.

O'Halloran, K.L. and Smith, B.A. (eds.), *Multimodal Studies: Exploring Issues and Domains*, Routledge, London New York 2011.

O'Neill, C., *Weapons of Math Destruction*, Penguin, London 2016.

Oppenlaender, J. *et al.*, “Prompting AI Art: An Investigation into the Creative Skill of Prompt Engineering,” in «International Journal of Human–Computer Interaction», 2024, pp. 1-23.

Osler, L., *Taking Empathy Online*, Routledge, London 2021.

Osoba, O.A. and Welser IV, W., *An Intelligence in Our Image The Risks of Bias and Errors in Artificial Intelligence*, RAND Corporation, Santa Monica 2017.

Ovid, *Metamorphoses,* eds. by F.J. Miller, G.P. Goold, Cambridge 1916.

Paananen, V. *et al.*, “Using Text-to-Image Generation for Architectural Design Ideation,” in «International Journal of Architectural Computing», 22, 3, 2023, pp. 458-74.

Paglen, T., "Invisible Images (Your Pictures are Looking at You)," in «The New Inquiry», 2016: https://thenewinquiry.com/invisible-images-your-pictures-are-looking-at-you/.

Paić, Ž. (ed.), *The Technosphere as a New Aestetic*, Cambridge Scholar Publishing, NewCastle 2022.

Paić, Ž. and Purgar, K. (eds.), *Theorizing Images*, Cambridge Scolars Publishing, Newcastle 2016.

Panofsky, E., *Architecture gothique et pensée scolastique*, French transl. and postface by P. Bourdieu, «Le sens commun», Éditions de Minuit, Paris 2016.

Panofsky, E., *Perspective as Symbolic Form* (1927), Engl. transl. by C. Wood, Zone Books, New York 1991.

Pariser, E., *The Filter Bubble: What the Internet is Hiding from You*, Penguin, London 2012.

Pasquale, F., *The Black Box Society*, Harvard University Press, Cambridge Mass. 2020.

Perone, U., "Il nome e l'immagine: la differenza dell'identità," in «Teoria», 1, 2010, pp. 72-3.

Pflanzer, M. *et al.*, "Ethics in human-AI teaming: principles and perspectives," in «AI Ethics», 3, 2023, pp. 917-35.

Pico della Mirandola, G., *L'immaginazione*, edited by F. Molinarolo, Edizioni della Normale, Pisa 2022.

Pinotti, A., and Somaini, A., *Cultura visuale. Immagini, sguardi, media e dispositivi*, Einaudi, Torino 2016.

Pinotti, A., "Staying Here, Being There. Bilocation, Empathy and Self-empathy in Virtual Reality," in «Philosophical Bulletin», 37, 2022, pp. 147-62.

Plato, *Protagoras*, Engl. transl. by B. Jowett and M. Ostwald, The Library of Liberal Arts, Bobbs-Merrill, Indianapolis – New York 1956.

Plato, *Republic*, ed. by R. Waterfield, Oxford University Press, Oxford 2008.

Plato, *Timaeus; and, Critias*, ed. by D. Lee, Penguin, London 1977.

Plotinus, *The Enneads*, ed. and trans by Lloyd P. Gerson, Cambridge University Press, Cambridge 2017.

Prunkl, C., "Human Autonomy at Risk? An Analysis of the Challenges from AI," in «Minds & Machines», 34, 26, 2024.

Prunkl, C., "Human Autonomy in the Age of Artificial Intelligence," in «Nature Machine Intelligence», 4, 2, 2022, pp. 99-10.

Qiao, S. *et al.*, "Reasoning with Language Model Prompting: A Survey," in «Proceedings of the 61st Annual Meeting of the Association for

Computational Linguistics», vol. 1, *Association for Computational Linguistics*, Toronto 2023, pp. 5368-93.

Raney, A., "The Role of Morality in Emotional Reactions to and Enjoyment of Media Entertainment," in «Journal of Media Psychology», 23, 1, 2011, pp. 18-23.

Reath, A., *Kant's Conception of Autonomy*, in O. Sensen (ed.), *Kant on Moral Autonomy*, Cambridge University Press, Cambridge 2013, pp. 32-52.

Reilly J., *et al.*, *What We mean When We Say Semantic: Toward a Multidisciplinary Semantic Glossary*, in «Psychonomic Bulletin & Review», 32, 2025, pp. 243-280.

Resnik, D., "Openness in Scientific Research: A Historical and Philosophical Perspective", in «Open Access Law», 21, 11, 1, 2023.

Restuccia, F., *Immaginazione artificiale? Flusser e l'intelligenza artificiale generativa*, in A. Fabris (ed.), *Immagini, media, dispositivi. Studi e prospettive*, Edizioni ETS, Pisa 2023.

Restuccia, F., *Il contrattacco delle immagini, tecnica, media e idolatria da Vilém Flusser*, Meltemi, Milano 2021.

Reijula, S. and Hertwig, R., "Self-nudging and the Citizen Choice Architect," in «Behavioural Public Policy», 2020, pp.1-30.

Ricoeur, P., "Les directions de la recherche philosophique sur l'imagination", in «Recherches phénoménologiques sur l'imaginaire I», Centre de Recherches Phénoménologiques, 29, Paris, 1974, pp. 1-8.

Ricœur, P., "La Bible et l'imagination," in «Révue d'Histoire et de Philosophie religieuses», 62, 1982.

Ricœur, P., *The Course of Recognition*, Engl. transl. by D. Pellauer, Harvard University Press, Harvard 2007.

Ricœur, P., "The Metaphorical Process as Cognition Imagination and Feeling," in «Critical Inquiry», 5, 1, 1978, pp. 143-59.

Rodolfi, A., *Introduzione*, in G. Garfagnini, A. Rodolfi (eds.), *Profezia, filosofia e prassi politica*, Edizioni ETS, Pisa 2013, pp. 7-17.

Rorty, R., *Contingency, Irony, and Solidarity*, Cambridge University Press, Cambridge 1989.

Rorty, R., *Introduction. Metaphilosophical Difficulties of Linguistic Philosophy,* in Id. (ed.), *The Linguistic Turn: Recent Essays in Philosophical Method*, University of Chicago Press, Chicago 1967.

Rosenblatt, F., "The Perceptron. A Perceiving and Recognizing Automaton", report no. 85-460-61, Cornell Aeronautical Laboratory Inc., January 1957.

Rozenberg, S., "From Continuous to Discrete to Continuous – Text-to-Image Models as Limit to Indeterminate Phantasy", in «Technophany», 3, 2, 2024, pp. 1-13.

Rumiati, R. and Legrenzi, P., "Euristiche: da Simon a Simon," in «Sistemi intelligenti, Rivista quadrimestrale di scienze cognitive e di intelligenza artificiale», 1, 2016, pp. 169-180.

Russel, S.J. and Norvig, P., *Artificial Intelligence. A Modern Approach*, Pertinence Hall, Upper Saddle River (NJ) 2021.

Samuel, A., "Some Studies in Machine Learning Using the Game of Checkers," in «IBM Journal of Research and Development», 3, 3, 1959, pp. 210-29.

Sandvig, C., "Corrupt Personalization", in «Social Media Collective», 26/6/2014.

Sanseviero, O. *et al.*, *Hands-On Generative AI with Transformers and Diffusion Models*, O'Reilly Media, Sebastopol 2024.

Sapiro, G., *Habitus*, in Ead. (dir.), *Dictionnaire international Bourdieu*, CNRS Éditions, Paris 2020, p. 387 ff.

Sartre, J.-P., *L'imagination* (1936), PUF, Paris 1989.

Sartre, J.-P., *L'Imaginaire. Psychologie phénoménologique de l'imagination*, Gallimard, Paris 1986.

Schick, N., *Deepfakes. The Coming Infocalypse*, Twelve, New York 2020.

Schmidt, E. *et al.*, *The Age of AI: And Our Human Future*, Little Brown and Company, Boston 2021.

Schneewind, J., *The Invention of Autonomy. A History of Modern Moral Philosophy*, Cambridge University Press, Cambridge 2010.

Schroeder, S.R., "Usage of Stereotypy as a Descriptive Term," in «Psychological Record», 20, 1970, pp. 337-342.

Schwartz, R. *et al.*, *Towards a Standard for Identifying and Managing Bias in Artificial Intelligence*, National Institute of Standards and Technology, Gaithersburg, MD 2022.

Seguró Mendlewicz, M., *On Vulnerability: A Philosophical Anthropology*, Lexington Books, Lanham 2024.

Sennett, R., *The Performer. Art, Life, Politics*, Allen Lane, London 2024.

Shah, H., "Algorithmic accountability," in «Philosophical Transactions: Mathematical, Physical and Engineering Sciences», 376, 2128, 2018.

Shawn, M. and Smithand, S. (ed.), *Photography and the Optical Unconscious*, Duke University Press, Durham and London 2017.

Simon, H.E., "Invariants of Human Behavior," in «Annual Reviews Psychology», 41, 1990, pp. 1-19.

Simon, H.E., "A Behavioral Model of Rational Choice," in «The Quarterly Journal of Economics», 69, 1, 1955, pp. 99-118.

Sinclair, A., "The Chameleon of Accountability: Forms and Discourses," in «Accounting, Organisations and Society», 20, 2, 1995, pp. 219-37.

Sisto, D., *Virtual influencer*, Einaudi, Torino 2024.
Slusher, M.P. and Anderson, C.A., "When Reality Monitoring Fails: The Role of Imagination in Stereotype Maintenance," in «Journal of Personality and Social Psychology», 52, 1987, pp. 653-62.
Somaini A., The "Medium of Perception: Walter Benjamin's Media Theory and the Tradition of The Media Diaphana", in Ž. Paić and K. Purgar (eds.), *Theorizing Images*, Cambridge Scolars Publishing, Newcastle 2016, pp. 84-110.
Sontang, S., *On Photography* (1977), Penguin, London 2008.
Spitzer, M., *Digitale Demenz*, Droemer, München 2012.
Spitzer, M., *Künstliche Intelligenz: Dem Menschen überlegen – wie KI uns rettet und bedroht,* Droemer, Munich 2023.
Sterelny, K., "Minds: Extended or Scaffolded?", in «Phenomenology and the Cognitive Sciences», 9, 4, 2010, pp. 465-81.
Straehle, C. (ed.), *Vulnerability, Autonomy, and Applied Ethics*, Routledge, London 2016.
Strawson, P.F., *Freedom and Resentment and Other Essays* (1974), Routledge, London 2008.
Sturloni, G., *La comunicazione del rischio per la salute e per l'ambiente*, Mondadori, Milano 2018.
Sue, K. *et al.* (eds.), *Disinformation, Misinformation, and Fake News in Social Media: Emerging Research Challenges and Opportunities*, Springer, Cham 2020.
Sueur, A.L., "Designing Redress: Who Does it, How and Why?," in «Asia Pacific Law Review», 20, 1, 2012, pp. 17-44.
Sunstein, C.R. and Cass, R., "Algorithms, Correcting Biases," in «Social Research: An International Quarterly», 86, 2, 2019, pp. 499-511.
Taddei Elmi, G., and Contaldo, A. (eds.), *Intelligenza artificiale, AI Act-Regolamento (UE) 1689/2024. Il nuovo scenario giuridico europeo*, Pacini, Pisa 2024.
Tajfel, H., "Cognitive Aspects of Prejudice," in «Journal of Social Issues», 25, 4, 1969, pp. 79-97.
Taylor, C., *A Secular Age*, Harvard University Press, Cambridge Mass. 2007.
Taylor, C., *Consciousness*, in P.F. Secord (ed.), *Explaining Human Behavior: Consciousness, Human Action and Social Structure*, Sage, London 1982, pp. 38-51.
Taylor, C., *Modern Social Imaginaries*, Durham, London 2004.
Taylor, C., *Sources of the Self: The Making of the Modern Identity*, Harvard University Press, Harvard 1992.

Taylor, C., *The Ethics of Authenticity*, Harvard University Press, Cambridge Mass. 1991.

Taylor, C., *The Opening Arguments of the Phenomenology*, in A. MacIntyre (ed.), *Hegel: a Collection of Critical Essays*, Doubleday, New York 1972.

Taylor, C., *Varieties of Religion Today: William James Revisited*, Harvard University Press, Cambridge Mass. 2002.

Taylor, C., *What is Human Agency?* in Id., *Human Agency and Language. Philosophical Papers I*, Cambridge University Press, Cambridge 1985, pp. 15-44.

Taylor, W.L., "Cloze Procedure: A New Tool for Measuring Readability," in «Journalism Quarterly», 30, 4, 1953, pp. 415-33.

Thaler R.H. and Sunstein C.R., *Nudge: Improving Decisions about Health, Wealth, and Happiness*, Penguin Books, London 2009.

Thapliyal, V. and Thapliyal, P., "AI and Creativity: Exploring the Intersection of Machine Learning and Artistic Creation," in «International Journal for Research Publication and Seminar», 15,1, 2024, pp. 36-41.

Tiribelli, S., *Identità personale e algoritmi. Una questione di filosofia morale*, Carocci, Roma 2023.

Tiribelli, S., *Moral Freedom in the Age of Artificial Intelligence*, Mimesis International, Milan-London 2022.

Toffler, A., *The Third Wave*, Bantam, New York 1991.

Tversky, C., *Mind in Motion: How Action Shapes Thought*, Basic Book, New York 2019.

Tversky, A. and Kahneman, D., "Availability: A Heuristic for Judging Frequency and Probability," in «Cognitive Psychology», 5, 1973, pp. 207-32.

Turati, F., *Atti del Parlamento italiano, Camera dei Seputati, sessioni 1904-1908, 17 giugno 1908*, n. 22962.

Turing, A.M., "Computing Machinery and Intelligence," in «Mind-New Series», 59, 236, 1950, pp. 433-60.

Uricchio, W., "The Algorithmic Turn: Photosynth, Augmented Reality and the State of the Image," in «Visual Studies», 1, 2011, pp. 25-35.

Usher, B., "Rethinking Microcelebrity: Key Points in Practice, Performance and Purpose", in «Celebrity Studies», 11, 2, 2020, pp. 171-88.

Vaccari, F., *Fotografia e inconscio tecnologico*, Einaudi, Torino 1979.

Van Es, K. and Nguyen, D., ""Your friendly AI assistant": the Anthropomorphic Self-representations of ChatGPT and its Implications for Imagining AI," in «AI & Society», 2024, pp. 3591-3603.

Varkey, B., "Principles of Clinical Ethics and Their Application to Practice," in «Medical Principles and Practice», 30, 1, 2021, pp. 17-28.

Vecchia, R., headword *Immaginazione*, in R. Romani (ed.), *Enciclopedia Einaudi*, vol. VII, Einaudi, Torino 1979, pp. 43-48

Vetlesen, A.J., *A Philosophy of Pain,* Reaktion Books, Edinburgh 2009.

Vico, G.B., *De antiquissima Italorum sapientia* (1710), edited by M. Sanna, Edizioni di Storia e Letteratura, Roma 2005.

Vincent, A., *A Structured Taxonomy of Responsibility Concepts*, in N.A. Vincent *et al.* (eds.), *Moral Responsibility. Beyond Free Will and Determinism*, Springer, Dordrecht 2011, pp. 15-35.

Virgil, *The Aeneid*, Engl. transl. by D. West, Penguin, London 2003.

Virilio, P., *L'Art à perte de vue*, Editions Galilée, Paris 2005, Engl. transl. by I. Rose, *Art as Far as the Eye Can See,* Berg Publishers, Oxford-New York 2007.

Von Weizsäcker, V., *Die Schmerzen* (1926), in Id., *Gesammelte Schriften 5*, Suhrkamp, Frankfurt am Main 1987.

Wahl, J., *La culture est-elle en péril? Débat sur ses moyens de diffusion: presse, cinéma, radio, télévision, Rencontres Internationales de Genève*, Editions de la Baconnière, Neuchâtel 1955, pp. 333-336

Warbug A., *Frammenti sull'espressione,* ed. by S. Müller, It. transl. by M. Ghelardi and G. Targia, Edizioni della Normale, Pisa 2011.

Warbug A., *Mnemosyne. Bilderatlas*, Dölling und Galitz Verlag, München-Hamburg 1994.

Warren, S. and Brandeis, L.D. "The Right to Privacy," in «Harvard Law Review», 193, 4, 1890, pp. 31, 33.

Webb, R., *Ekphrasis, Imagination and Persuasion in Ancient Rhetorical Theory and Practice*, Routledge, Farnham-Burlington 2009.

Wellner, G., *Can Algorithms Imagine?* in Ead. *et al.* (eds.),*The Philosophy of Imagination. Technology, Art and Ethics*, Bloomsbury, London 2024, pp. 30-41.

Wellner, G., "Digital Imagination, Fantasy, AI Art", in «Foundations of Science», 27, 4, 2022, pp. 1445-1451.

Westerlund, M., "The Emergence of Deepfake Technology: A Review," in «Technology Innovation Management Review», 9, 2019, pp. 40-53.

Wiedemann, J. (ed.), *Digital Beauties, 2D & 3D Computer Generated Digital Models, Virtual Idols and Characters*, Taschen, Koln-London-Madrid-New York-Paris-Tokyo 2001.

Wiesing, L., "Phänomenologie des Bildes nach Husserl und Sartre," in «Phänomenologische Forschungen», 30, 1996, pp. 255-81.

Wiesing, L., *Artifizielle Präsenz. Studien zur Philosophie des Bildes* (2005), Suhrkamp, Frankfurt am Main 2014.

Wittgenstein, L., *Philosophische Untersuchungen*, eds. by G.E.M. Anscombe and R. Rhees, Blackwell, Oxford 1953.

Wolf, M., *Reader. Come Home: The Reading Brain in a Digital World*, Harper Collins, New York 2018.

Wu, Y. *et al.*, "Image-Perfect Imperfections: Safety, Bias, and Authenticity in the Shadow of Text-To-Image Model Evolution," in «ACM Conference on Computer and Communications Security», October 14-18, 2024.

Wunenburger, J.-J., *La vie des images* (1995), Les Éditions Spéciales, Paris 2022.

Wunenburger, J.-J., "L'image, reproduction, transformation, création de l'"irréel" ? Quelques notes sur l'anthropologie de l'imagination ", in «IRIS», 44, 2024.

Wunenburger, J.-J., *Philosophies des images*, PUF, Paris 1997.

Xie, T. and Pentina, I., "Attachment theory as a framework to understand relationships with social chatbots: a case study of Replika", in «Proceedings of the 55th Hawaii International Conference on System Sciences», 2022, pp. 2046-55.

Yu, J., *et al.*, "Machine learning for big visual analysis," in «Machine Vision and Applications», 29, 2018, pp. 929-31.

Yuan, Z. *et al.*, "Protecting Copyright of Stable Diffusion Models from Ambiguity Attacks," in «Signal Processing», 227, 13, 2025, p.109722.

Zhao, Y., "GAN-Based Image Generation," in «Applied and Computational Engineering», 92, 2024, pp. 128-135.

Zillmann, D., *Empathy: Affective Reactivity to Others' Emotional Experiences*, in J. Bryant and P. Vorderer (eds.), *Psychology of Entertainment,* Lawrence Erlbaum Associates Publishers, Mahwah 2006, pp. 151-181.

Zowghi, D. and Bano, M., "AI for all: Diversity and Inclusion in AI," in «AI Ethics», 4, 2024, pp. 873-6.

Zuboff, S., *The Age of Surveillance Capitalism. The Fight for a Human Future at the New Frontier of Power*, Public Affairs, New York 2019.

Zylinska, J., *The Perception Machine. Our Photographic Future between the Eye and AI*, MIT Press, Boston 2023.

MIMESIS GROUP
www.mimesis-group.com

MIMESIS INTERNATIONAL
www.mimesisinternational.com
info@mimesisinternational.com

MIMESIS EDIZIONI
www.mimesisedizioni.it
mimesis@mimesisedizioni.it

ÉDITIONS MIMÉSIS
www.editionsmimesis.fr
info@editionsmimesis.fr

MIMESIS COMMUNICATION
www.mim-c.net

MIMESIS EU
www.mim-eu.com

www.ingramcontent.com/pod-product-compliance
Lightning Source LLC
Chambersburg PA
CBHW031344230926
51848CB00002B/379
* 9 7 8 8 8 6 9 7 7 5 0 0 0 *